The Molecular Life of Plants

The Molecular Life of Plants

Editors

Sangeeta Karnik, M. M. Sawant

scitus
academics

The Molecular Life of Plants

Edited by **Sangeeta Karnik, M. M. Sawant**

Printed in 2017

ISBN: 978-1-68117-101-2

Library of Congress Control Number: 2015935501

SCITUS Academics LLC,
616, Corporate Way, Suite 2, 4766,
Valley Cottage, NY 10989

www.scitusacademics.com

Notice

Reasonable efforts have been made to publish reliable data and views articulated in the chapters are those of the individual contributors, and not necessarily those of the editors or publishers. Editors or publishers are not responsible for the accuracy of the information in the published chapters or consequences of their use. The publisher believes no responsibility for any damage or grievance to the persons or property arising out of the use of any materials, instructions, methods or thoughts in the book. The editors and the publisher have attempted to trace the copyright holders of all material reproduced in this publication and apologize to copyright holders if permission has not been obtained. If any copyright holder has not been acknowledged, please write to us so we may rectify.

Contents

Preface

The Molecular Life of Plants presents students with an innovative, integrated approach to plant science. It looks at the processes and mechanisms that underlie each stage of plant life and describes the intricate network of cellular, molecular, biochemical and physiological events through which plants make life on land possible. Richly illustrated, this book follows the life of the plant, starting with the seed, progressing through germination to the seedling and mature plant, and ending with reproduction and senescence. This "seed-to-seed" approach will provide students with a logical framework for acquiring the knowledge needed to fully understand plant growth and development. This book will prove invaluable to students needing a comprehensive, integrated introduction to the subject across a variety of disciplines including plant science, biological science, horticulture and agriculture.

Editor

Chapter 1

The Molecular Mechanism of Sporocyteless/Nozzle in Controlling Arabidopsis Ovule Development

Baoye Wei[1], Jinzhe Zhang[1], Changxu Pang[1], Hao Yu[1], Dongshu Guo[1], Hao Jiang[1], Mingxin Ding[1], Zhuoyao Chen[1], Qing Tao[1], Hongya Gu[1,2], Li-Jia Qu[1,2,3], and Genji Qin[1]

[1]State Key Laboratory of Protein and Plant Gene Research, College of Life Sciences, Peking University, Beijing 100871, China

[2]The National Plant Gene Research Center (Beijing), Beijing 100101, China

[3]Peking-Tsinghua Center for Life Sciences, Peking University, Beijing 100871, China

ABSTRACT

Ovules are essential for plant reproduction and develop into seeds after fertilization. *SPOROCYTELESS/NOZZLE* (*SPL/NZZ*) has been known for more than 15 years as an essential factor for ovule development in *Arabidopsis*, but the biochemical nature of SPL function has remained unsolved. Here, we demonstrate that SPL functions as an adaptor-like transcriptional repressor. We show that SPL recruits TOPLESS/TOPLESS-RELATED (TPL/TPR) co-repressors to inhibit the CINCINNATA (CIN)-like TEOSINTE BRANCHED1/CYCLOIDEA/PCF (TCP) transcription factors. We reveal that SPL uses its EAR motif at the C-terminal end to recruit TPL/TPRs and its N-terminal part to bind and inhibit the TCPs. We demonstrate that either disruption of *TPL/TPRs* or overexpression of *TCPs* partially phenocopies the defects of megasporogenesis in *spl*. Moreover, disruption of *TCP*s causes phenotypes that resemble *spl-D* gain-of-function mutants. These results define the action mechanism for SPL, which along with TPL/TPRs controls ovule development by repressing the activities of key transcription factors. Our findings suggest that a similar gene repression strategy is employed by both plants and fungi to control sporogenesis.

INTRODUCTION

In flowering plant female reproduction, an ovule consists of the haploid female gametophyte and the diploid integuments and develops into a seed after fertilization. The first step of ovule development is the formation of a protrusion from the internal wall of the carpel. The protrusion then elongates and forms a finger-like nucellus. At the tip of the nucellus, only one hypodermal cell differentiates into an archespore, which further elongates and differentiates into a megasporocyte. A megasporocyte goes through meiosis to produce four haploid megaspores. In *Arabidopsis*, three of the megaspores near the distal end of the ovule are degenerated and the remaining cell develops into the haploid embryo sac by undergoing three rounds of mitosis. In the mature ovule, the embryo sac contains one egg cell, one central cell, two synergids and three antipodal cells. Extensive genetic studies in *Arabidopsis* have identified several components that play key roles in ovule development (see reviews[1,2,3,4]). Among them, SPOROCYTELESS/NOZZLE (SPL/NZZ) is a key regulator responsible for promoting the formation of megasporocyte and integuments during ovule development[5,6,7,8].

The *spl/nzz* mutants were initially identified as sterile mutants[5,6], which displayed a complete failure of male and female gametophyte formation. Detailed analysis showed that SPL/NZZ controlled sporogenesis and played pivotal roles in the differentiation of megasporocytes. In the embryo sac of *spl/nzz* mutants, an archesprorial cell is formed but it fails to differentiate into a megasporocyte. SPL/NZZ also promotes the growth of ovule integuments[8], the differentiation of microsporocytes[9], the identity of stamen[10], and other lateral organ development[11]. It was reported that SPL/NZZ physically interacted with the YABBY protein INNER NO OUTER (INO) in the regulation of ovule development[12]. Some important downstream genes of SPL/NZZ have also been identified. During ovule development, SPL/NZZ represses some important ovule developmental genes including *BELL1* (*BEL1*), *AINTEGUMENTA* (*ANT*) and *INO*[12], and the auxin biosynthesis gene *YUCCA2*[11,13]. It has been shown that AGAMOUS activates the expression of *SPL/NZZ* by directly binding to the CArG box at the 3′-region of the gene during microsporogenesis[9]. Recently, the plant hormone cytokinin has been reported to promote the expression of *SPL/NZZ*, and SPL/NZZ in turn activates the auxin transport gene *PIN1* during ovule development[14]. However, the exact action mode of SPL/NZZ is still not understood.

We previously isolated a dominant *spl-D* mutant from an *Arabidopsis* activation tagged mutant collection and found that SPL controls lateral organ morphogenesis by regulating auxin homeostasis[11]. Here, we demonstrate that SPL/NZZ is actually a transcriptional repressor. The C-terminal end of SPL/NZZ contains a typical EAR motif (ERF-associated amphiphilic repression), which is a well-characterized repression domain with the consensus sequence LXLXL[15,16]. We show that SPL/NZZ uses its EAR motif to recruit the known transcriptional co-repressor TOPLESS/TOPLESS-RELATED (TPL/TPRs). We demonstrate that *TPL* and *SPL* have overlapping expression patterns and that the homozygous mutants *tpl-1* and *spl*displayed similar defects in ovule development. Furthermore, we discover that SPL/NZZ, through its N-terminal part without the EAR motif, directly interacts with the CINCINNATA (CIN)-like TCP transcription factors. Disruption of CIN-like TCPs leads to distorted ovule arrangement in ovaries similar to those observed in the gain-of-function mutant *spl-D*. Overexpression of the CIN-like *TCPs* resulted in aborted ovules similar to those observed in the *spl* loss-of-function mutant. Our results indicate that SPL serves as an adaptor-like transcriptional repressor to control ovule development by recruiting TPL/TPR co-repressors to suppress the activities of CIN-like TCP transcription factors in *Arabidopsis*. Interestingly, both plants and fungi adopt a similar gene repression strategy to control sporogenesis, suggesting that this strategy may be an evolutionarily conserved mechanism.

RESULTS

SPL/NZZ is an EAR Motif-containing Transcriptional Repressor

SPL/NZZ was reported as a nuclear protein with several structural features including a basic region in the N-terminus and two putative helixes[5]. We re-analyzed the protein sequence of SPL and found that the last five amino acid residues at the C-terminal end, i.e., LSLKL, resembled the EAR motif (LXLXL;Figure 1A). Because many known repressors including IAA12, NINJIA, and TIE1 all contain the EAR motif[17,18,19] (Figure 1B), SPL may also have transcriptional repression activity. To test this hypothesis, we co-transformed GAL4 DNA binding domain (G4DBD) or G4DBD-SPL with a reporter construct 35S-UAS-GUS[19] into tobacco leaves. The results showed that GUS activities were highly repressed in the combination of 35S-UAS-GUS and G4DBD-SPL, suggesting that SPL has transcriptional repression activities (Figure 1C). To further determine whether the EAR motif in SPL is required for the repression activity, we generated two other fusion constructs, G4DBD-SPLΔEAR in which the EAR motif was deleted and G4DBD-SPLmEAR in which the conserved leucine residues in the EAR motif were mutated to alanine residues. We found that the GUS activities were not repressed in the tobacco leaves in the combinations of the reporter and either G4DBD-SPLΔEAR or G4DBD-SPLmEAR (Figure 1C), indicating that the EAR motif is required for the repression activity of SPL.

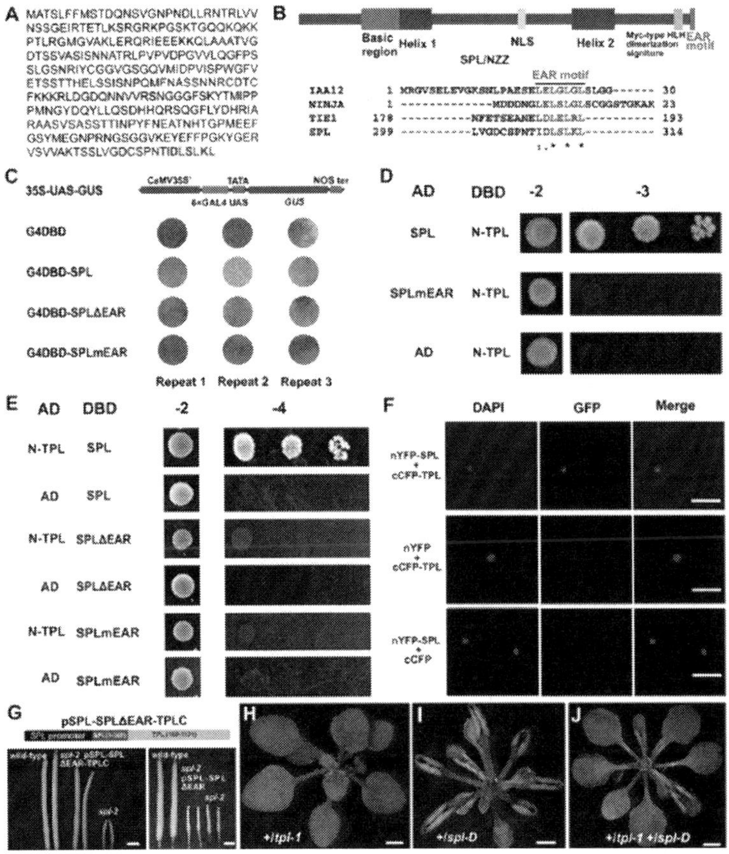

Figure 1: SPL/NZZ is a transcriptional repressor interacting with the TPL co-repressor. (A, B) SPL contains a typical EAR motif. The full-length protein sequence of SPL. The EAR motif at the C-terminal end is indicated by color letters (A). The domain structures of SPL are represented by different color blocks. The EAR motif is shared by other known repressors (B). (C) SPL shows transcriptional repression activity dependent on the C-terminal EAR motif. The transcription activity of SPL was investigated in tobacco leaves. The small circles were tobacco leaf discs co-transformed with different construct combinations. Every combination was repeated for three times. (D, E) Yeast two-hybrid assays indicated that SPL interacted with TPL protein by the EAR motif. Transformed yeast cells were spotted on control medium (2), or selective medium SD-Leu-Trp-His (−3), or SD-Ade-Leu-Trp-His (−4) in 10-fold, 100-fold and 1 000-fold dilutions. The empty vectors were used as the controls. (F) BiFC analysis verified the interaction of SPL and TPL. From left to right, DAPI staining of the nucleus, GFP fluorescence, and merge of DAPI and GFP.

Scale bar, 30 μm. (G) pSPL-SPLΔEAR-TPLC fusion protein complemented the male and female sterility of *spl-2*. Top, schematic representation of pSPL-SPLΔEAR-TPLC. Bottom left, the siliques from wild-type, *spl-2*transformed with pSPL-SPLΔEAR-TPLC and *spl-2*. Bottom right, pSPL-SPLΔEAR could not complement the male and female sterility of *spl-2*. (H-J) The genetic interaction of *spl-D* and *tpl-1*. The up-curled leaf phenotype of heterozygous *spl-D* was released in the background of *tpl-1* mutant. 21-day-old heterozygous *tpl-1*mutant (G). 21-day-old heterozygous *spl-D* mutant (H). 21-day-old heterozygous *tpl-1 spl-D* double mutant (I). Scale bar, 1 mm.

SPL/NZZ Interacts with TPL/TPR Co-repressors

EAR motif-containing repressors often interact with TPL/TPR co-repressors[17,18,19]. To test whether SPL could interact with TPL/TPRs, we fused GAL4 activation domain (AD) to SPL or SPLmEAR, and DNA binding domain (DBD) to the N-terminal region of TPL (N-TPL), a region sufficient for interaction of TPL with EAR motif-containing proteins[17,20], to conduct the yeast two-hybrid (YTH) assays. The results showed that SPL interacted with TPL, whereas SPLmEAR did not (Figure 1D). We further fused DBD to either SPL, SPLmEAR, or SPLΔEAR, and AD to the N-TPL, and the YTH results confirmed that SPL interacted with TPL and that the EAR motif was required for the interaction (Figure 1E). Bimolecular fluorescence complementation (BiFC) assays also confirmed the interaction between SPL and TPL *in vivo* (Figure 1F). Furthermore, SPL also interacted with the N-terminus of the other four TPRs (Supplementary information, Figure S1).

To further test the hypothesis that SPL regulates plant development by recruiting TPL through the EAR motif, we first identified a 4.4-kb-long fragment of the *SPL*promoter, pSPL, which reproduced the expression pattern of *SPL* (Supplementary information, Figure S2A and S2B)[9]. We used the SPL promoter to drive SPLΔEAR-TPLC in which SPLΔEAR fused with the C-terminus of TPL (TPLC) to generate pSPL-SPLΔEAR-TPLC (Figure 1G). We then identified *SAIL_519_H07* T-DNA insertion mutant as *spl-2*. The *spl-2* contained a T-DNA insertion located 206 bp upstream to the start codon ATG of *SPL* (Supplementary information, Figure S2C) and displayed male and female sterility (Supplementary information, Figure S2D and S2E, Figure 1G). We transformed pSPL-SPLΔEAR-TPLC or the control pSPL-SPLΔEAR into *spl-2*. It was clear that pSPL-SPLΔEAR-TPLC complemented the male and female sterility of *spl-2* (Figure 1G), whereas pSPL-SPLΔEAR did not (Figure 1G), suggesting that the SPLΔEAR-TPLC can functionally substitute SPL. To further demonstrate the essential roles of EAR motif for SPL function, we generated 35S-SPLmEAR plants in which SPLmEAR was driven by a CaMV 35S promoter (Supplementary information, Figure S2F). We hypothesized that SPLmEAR could compete with endogenous SPL, thus causing a dominant-negative effect. Indeed, overexpression of SPLmEAR led to male and female sterility, resembling *spl-2* (Supplementary information, Figure S2D-S2H; Figure 1G), indicating that the EAR motif was crucial for SPL functions. We also observed epinastic leaves in the 35S-SPLmEAR plants, a phenotype opposite to the hyponastic leaves in *spl-D* mutants, suggesting that SPLmEAR may also compete with and disrupt the functions of some other unknown proteins redundant to SPL in leaves (Supplementary information, Figure S2I-S2L). When we crossed *spl-D*with a dominant-negative mutant *tpl-1*[17], we found that the hyponastic leaf phenotype of *+/spl-D* was partially rescued in the *+/spl-D +/tpl-1*, suggesting that TPL is required for the SPL function[11] (Figure 1H-1J). These results demonstrate that SPL represses transcription by interacting with TPL/TPRs through the EAR motif.

TPL is expressed during Ovule Development

To test whether TPL participates in ovule development, we first analyzed the expression pattern of *TPL* in the ovules using the TPLP-GUS plants[19]. *TPL* was expressed ubiquitously in the pistils, stamens and pollens (Figure 2A, 2B, 2G, and2H), and expressed dynamically during ovule development. GUS signal was first relatively strong in the proximal region and weak in the nucellus in the young ovules (Figure 2C). Then GUS signal became stronger and more ubiquitously distributed in both proximal region and nucellus (Figure 2D and 2E). In the mature ovules, GUS staining was mainly observed in the nucellus (Figure 2F). This expression pattern of *TPL* is consistent with that revealed by *in situ*hybridizations[21]. Quantitative real-time PCR (qRT-PCR) analysis confirmed that the expression of *SPL* and *TPL* were overlapped (Figure 2I). These results suggest that TPL may be involved in ovule development.

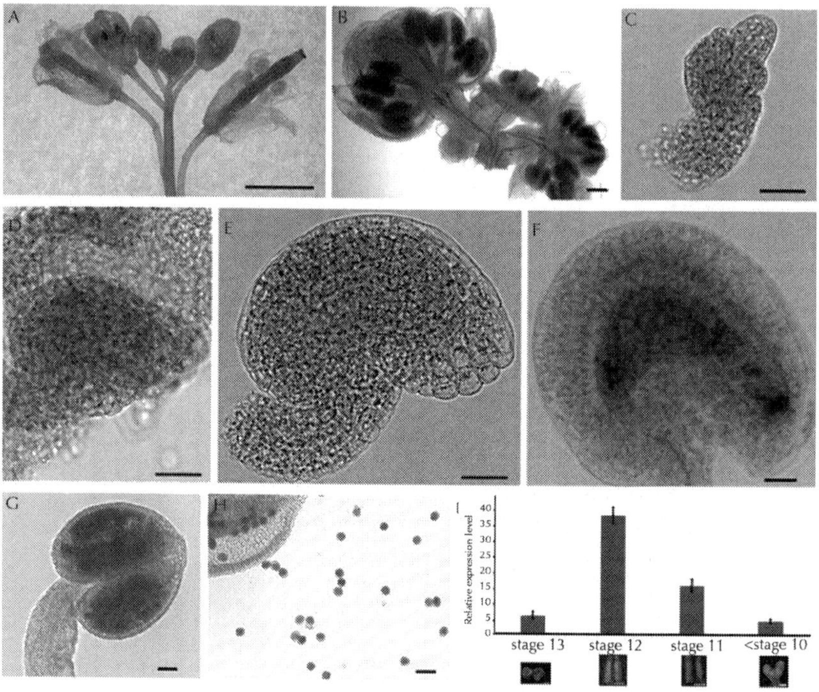

Figure 2: The expression pattern of *TPL*. (A, B) GUS staining of inflorescences (C-E) *TPL* was expressed in young ovules. (F) *TPL*was expressed strongly in the nucellus of a mature ovule. (G, H)*TPL* was highly expressed in an anther and pollens. (I) The overlapping expression of *TPL* and *SPL* revealed by qRT-PCR. The relative expression levels of *TPL* in the mature ovules from flower stage 13, the pistils from flower stage 12 or flower stage 11, and the flower buds before stage 10 The expression level of *SPL* was set to 1.0 in the different corresponding tissues. The error bars represent the SD of three biological replicates. Scale bars, 1 mm (A), 100 μm (B), 20 μm (C-F), and 50 μm (G, H).

TPL/TPR Co-repressors are required for Ovule Development

To reveal the possible roles of TPL/TPR co-repressors during ovule development, we quantitatively and statistically analyze the phenotypes of *tpl-1*, a dominant-negative mutant of *TPL/TPRs*[22]. We grew and identified 1 638 homozygous *tpl-1*and found 1 210 plants with cup-shaped cotyledons, 372 with one cotyledon, 5 with no cotyledon, and

51 plants with two cotyledons which were similar to wild-type (WT) plants. Only 184 plants grew to adult stage and flowered. We dissected the siliques of 116 *tpl-1* individual plants and found variable aborted ovules (Figure 3I and 3J). We took one silique from each of 61 *tpl-1* plants and 10 WT plants, and observed the ovules. In WT siliques only 3.1% of the ovules (17/550) were aborted, whereas 43.8% (550/1 254) of ovules in *tpl-1* siliques were aborted, suggesting that *TPL/TPRs* are required for ovule development.

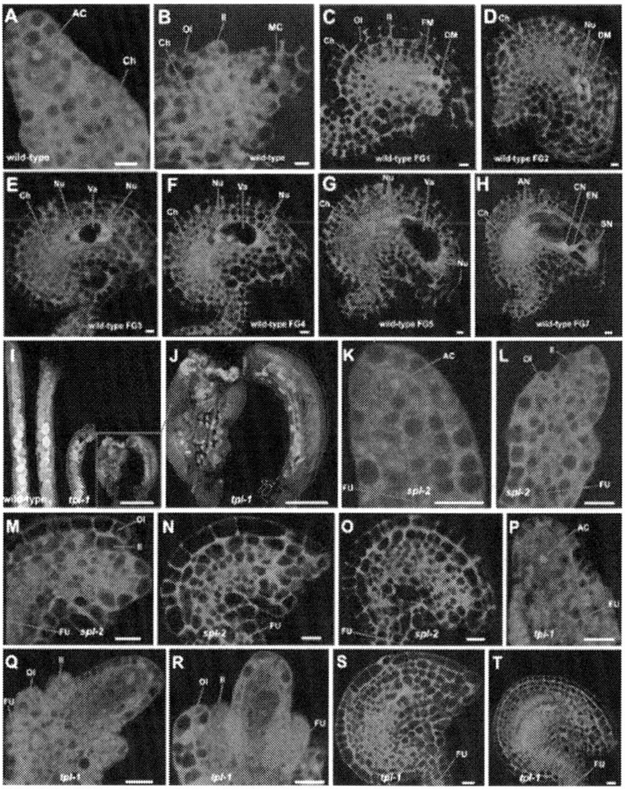

Figure 3: The developmental ovules from wild type, *spl-2* and *tpl-1*observed by confocal laser scanning microscopy (CLSM). (A-H)The developmental ovules from wild type observed by CLSM. An archesporial cell was formed in the primordium (A). A megasporocyte could be found (B). Functional and degenerating megaspores were observed (C). Vacuole formation, nuclear division, nuclear migration and fusion, and cellularization could be found (D-H). Scale bar, 5 μm. (I, J) The dissected siliques from wild-type and four

tpl-1 individual plants. Aborted ovules were prevalent in *tpl-1* siliques. From left to right, silique from wild-type plant, four siliques from *tpl-1* plants (I). The close-up view of aborted ovules in *tpl-1* siliques (J). Scale bars, 1 mm (I) and 500 μm (J). (K-O) The ovules from *spl-2* mutant. An archesporial cell was formed in the ovule of *spl-2* (K). No megasporocyte could be found (L). No events, such as vacuole formation, nuclear division, and cellularization could be found (M-O). (P-T) The ovules from*tpl-1* plants. An archesporial cell was also observed (P). No megasporocytes or functional megaspores could be found in the ovules (Q, R). And no nuclei, vacuoles, egg cell, central cell or synergid cells could be found in the ovules (S, T). Scale bar, 10 μm (K-T). AC, Archesporial cell; MC, Megasporocyte; OI, Outer integument; II, Inner integument; FM, Functional megaspore; DM, Degenerating megaspore; Nu, Nucleus; Va, Vacuole; AN, Antipodal cell nucleus; CN, Central cell nucleus; EN, Egg cell nucleus; SN, Synergid cell nucleus; FU, Funiculus.

We further adopted confocal laser scanning microscopy (CLSM) to analyze the ovules of *spl-2* and *tpl-1* mutants. In the WT ovules, an archesporial cell was formed in the primordium (Figure 3A) and then differentiated into a megasporocyte (Figure 3B). The megasporocyte underwent meiosis, and one of the meiotic products developed into a functional megaspore (Figure 3C). Then the megaspore produced a two-nucleated embryo sac (Figure 3D and 3E). After a series of events, i.e., vacuole formation, two other rounds of nuclear division, nuclear migration and fusion, and cellularization (Figure 3E, 3F, 3G, and 3H), a seven-celled embryo sac containing one egg cell, one central cell, two synergid cells and three antipodal cells was formed (Figure 3H). However, in the *spl-2* and *tpl-1*mutants showing low fertility, although the archesporial cells were formed in nearly all the observed ovules (Figure 3K and 3P), no megasporocytes were found in 41.8% (69/165) of *tpl-1* ovules, a phenotype similar to that of *spl-2* (Figure 3Land 3Q). In the ovules of *spl-2* and *tpl-1* without megasporocytes, vacuole formation, nuclear division and cellularization were not observed (Figure 3M, 3N, 3R, and 3S). Both *spl-2* and *tpl-1* produced ovules without egg cells, central cells, synergid cells and antipodal cells (Figure 3O and 3T). The development of *tpl-1*ovules with megasporocytes was frequently arrested at different stages, indicating that TPL/TPRs may also be essential for megagametogenesis (Supplementary information, Figure S3A-S3F). The *tpl-1* mutants also produced defective pollens (Supplementary information, Figure S3G and S3H). The observation that *spl-2* and *tpl-1* displayed similar ovule defects further supports our hypothesis that both SPL and TPL are involved in ovule development

SPL Interacted with CIN-like TCP Transcription Factors

We conducted a yeast two-hybrid screen using SPL as a bait to identify proteins that interact with SPL[23]. Several classes of transcription factors were identified from the screen, including the CIN-like TCPs (Supplementary information, Figure S4A) and the known SPL-interacting protein INO[12]. We further verified that all eight CIN-like TCPs interact with SPL in yeast cells (Figure 4A). Moreover, we demonstrated that CIN-like TCPs interacted with the N-terminal part of SPL without the EAR motif (Supplementary information, Figure S4B). Firefly luciferase complementation imaging assays also confirmed that SPL interacted with TCP2, TCP3, TCP5, TCP10, TCP17, or TCP24 *in vivo* (Figure 4B and 4C; Supplementary information, Figure S4C-S4F). The fact that SPL interacts with TPL/TPR co-repressors with its C-terminal EAR motif and interacts with TCPs with its N-terminal part raises the possibility that SPL serves as a bridge to link TPL/TPRs to TCPs. To test the hypothesis, we first conducted qRT-PCR analysis and found that the TCP direct target genes were downregulated in the *SPL* gain-of-function mutant *spl-D* and in *jaw-5D* mutant in which the microRNA *miR319* was overexpressed to inhibit five CIN-like *TCP* genes[19,24], suggesting that overproduction of SPL could repress TCP activities (Supplementary information, Figure S5A)[11,25,26,27]. The repression of the TCP target genes was largely relieved by the disruption of TPL/TPRs (Supplementary information, Figure S5A). We then used a CaMV 35S promoter to drive TCP17 which was fused with the C-terminal half of SPL (Figure 4D). All of the 10 independent 35S-TCP17-SPLC transgenic plants, but not the controls, displayed up-curled cotyledons and produced unfurled flower buds that were also observed in *SPL* overexpression plants (Figure 4D and 4E; Supplementary information, Figure S5B, S5C and S5D)[10,11], confirming the association between SPL and TCPs. We then crossed*spl-D* with *jaw-5D*. The *jaw-5D +/spl-D* double mutants produced more highly serrated and more deeply lobed leaves than *jaw-5D* or *+/spl-D* (Supplementary information, Figure S5E-S5G), suggesting that *spl-D* synergistically interacted with*jaw-5D*. These results suggest that SPL connects CIN-like TCP transcription factors with TPL/TPR co-repressors.

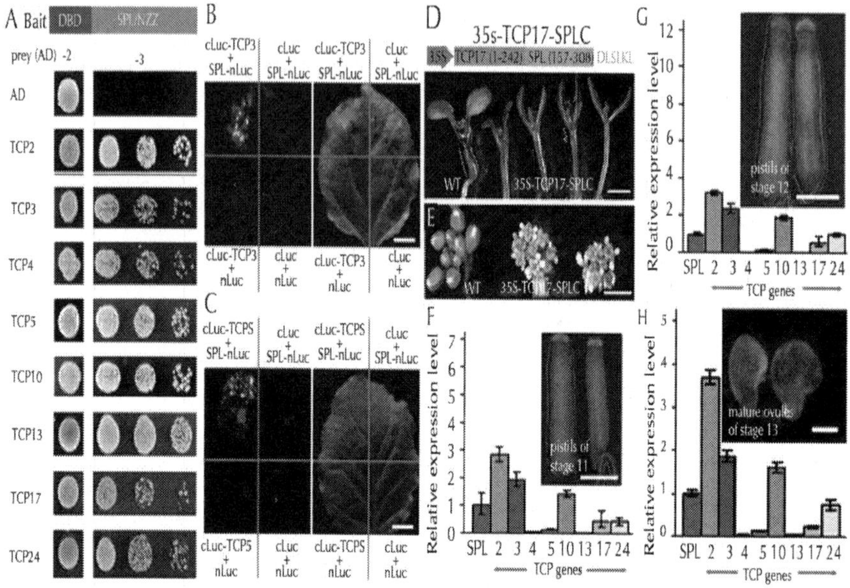

Figure 4: SPL interacts with CIN-like TCPs and most of them are expressed during ovule development. (A) SPL interacted with CIN-like TCP transcription factors in yeast cells. Transformed yeast cells were spotted on control medium (−2) or selective medium (−3) in 10-fold, 100-fold and 1 000-fold dilutions. The empty vector was used as the control. (B, C) SPL was confirmed to interact with TCP3 (B) or TCP5 (C) *in vivo* by firefly luciferase complementation imaging assays. (D, E) Overexpression of TCP17-SPLC fusion protein led to the up-curled cotyledons (D) and unfurled flower buds (E) similar to those of *SPL* overexpression lines. Top of D, schematic representation of 35S-TCP17-SPLC construct. (F-H) The relative expression levels of CIN-like *TCPs* in the pistils from stage 11 (F) and stage 12 (G), and the mature ovules from stage 13 (H) The expression level of *SPL* was set to 1.0. The error bars represent the SD of three biological replicates. Scale bars, 1 cm(B, C), 1 mm (D, E), 200 μm (F, G), 50 μm (H).

Most of CIN-like TCPs are expressed during Ovule Development

To test whether CIN-like TCPs play roles in ovule development, we first analyzed the expression of CIN-like *TCP* genes during ovule development. Ovules start initiation at flower stage 8. The megasporocytes are differentiated at stage 11 and the megagametogenesis begins at stage 12. Ovules are mature at stage 13 (Supplementary information, Figure S6A)[28]. We tested the expression of *SPL* in the flowers of different stages, i.e., flower buds before stage 10, the flowers of stage 11, stage 12 and stage 13 (Supplementary information, Figure S6B). The dynamic expression of *SPL* was consistent with the results of SPL-GUS staining[9]. We then tested the expression of the CIN-like *TCPs* and found that the CIN-like*TCP* genes were all expressed in these flowers (Supplementary information, Figure S6C-S6F). We next isolated the pistils from flowers at stage 11, stage 12, and the mature ovules from flower stage 13 for more precise analysis. qRT-PCR analysis showed that six *TCP* genes were expressed in these tissues, with high expression from *TCP2, TCP3* and *TCP10,* and low expression from *TCP5, TCP17* and*TCP24* (Figure 4F-4H), suggesting that CIN-like TCPs may participate in ovule development.

The Activities of CIN-like TCPs are Essential for Ovule Development

To investigate the possible roles of CIN-like TCPs during ovule development, we examined the ovule development of the four mutants, i.e., *tcp3 tcp4 tcp5 tcp10*quadruple mutant, *jaw-5D* in which *TCP2, TCP3, TCP4, TCP10* and *TCP24* were downregulated, *spl-D,* and *jaw-5D +/spl-D* double mutant. Interestingly, all these four mutants produced seeds that were protruded from siliques and the double mutant *jaw-5D +/spl-D* displayed enhanced phenotypes (Figure 5A). Scanning electron microscope analysis showed that the arrangement of ovules, which are normally interdigitated in WT ovaries (Figure 5B), were frequently disrupted in*tcp3 tcp4 tcp5 tcp10, jaw-5D* and *+/spl-D* mutants: two ovules arose consecutively from the same side of the locule (Figure 5C-5E). In *jaw-5D +/spl-D,* more ovules were initiated consecutively and crowded at one side of the locule (Figure 5F), indicating that disruption of the CIN-like TCPs caused the abnormal ovule initiation in the ovaries.

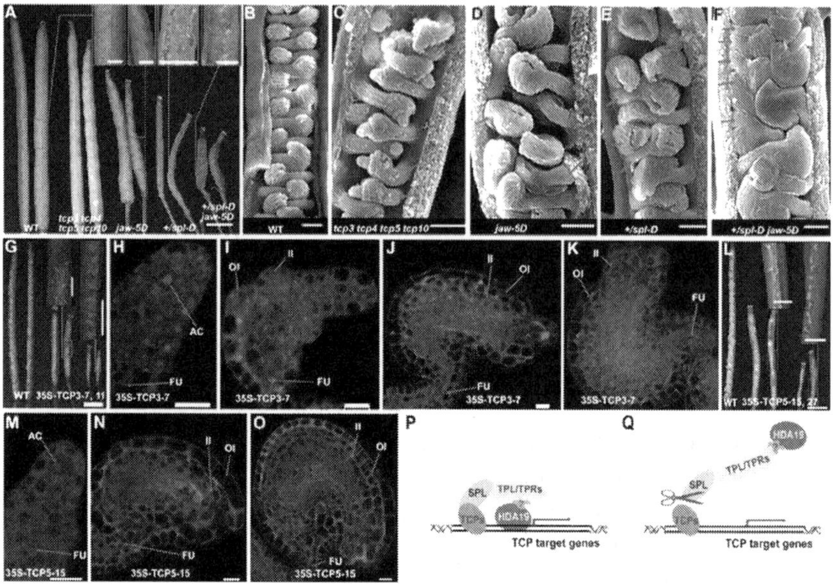

Figure 5: The CIN-like *TCPs* are implicated in ovule development. (A) The seeds are protruded from siliques from the CIN-like TCP disruption mutants and *spl-D*. Scale bars, 1 mm and 250 μm in the insets. (B-F) The interdigitated array of ovules was disrupted in the CIN-like TCP disruption mutants and *spl-D*. The interdigitated array of ovules in young pistil of wild type (B). The abnormal arrays of ovules in *tcp3 tcp4 tcp5 tcp10* (C), *jaw-5D* (D),*spl-D* (E) and *jaw-5D +/spl-D*(F). The red arrows indicate the consecutive ovules. Scale bar, 50 μm. (G-K) The*TCP3*overexpression lines produced aborted ovules. The dissected siliques from wild-type and two 35S-TCP3 plants (G). The ovule of 35S-TCP3 plant produced an archesporial cell (H). No megasporocyte was formed in the ovule (I). No egg cells, central cells, synergid cells and antipodal cells were formed in the ovules(J, K). The integuments of the ovules displayed retarded growth. Scale bars, 1 mm (G) and 250 μm in insets of G, and 10 μm (H-K). (L-O) Overexpression of *TCP5* caused aborted ovules similar to those in *spl* mutants. The dissected siliques showed high ratios of aborted ovules (L). The archesporial cell in the tip of the primordium contained a big nucleus (M). No megasporocytes or megaspores were observed in the ovule (N). The nucellus contained no egg cells, central cells, synergid cells and antipodal cells (O). Scale bars, 1 mm (L) and 250 μm in the insets of L and 10 μm (M-O). AC, Archesporial cell; FU, Funiculus; OI, Outer integument; II, Inner integument. (P, Q) The working model of SPL. (P) In the context of *SPL* expression, the N-terminus of SPL interacts with the CIN-like TCPs, while the EAR motif at the C-terminus interacts with TPL/TPR co-repressors. The

activities of TCPs are modified by the connection with TPL/TPRs. The TCP downstream genes are possibly regulated so that the differentiation of megasporocytes is promoted. (Q) When SPL is inactivated, TPL/TPRs are dissociated from CIN-like TCPs. The excessive activities of TCPs alter the expression of TCP downstream genes and cause abnormal ovule development.

We further used a CaMV 35S promoter to overexpress *TCP3* and *TCP5*, two representative members from the two main clades of CIN-like TCPs (Supplementary information, Figure S4A). All of the six independent 35S-TCP3 lines and nine of the 35S-TCP5 lines produced aborted ovules (Figure 5G and 5L). CLSM analysis showed that archesporial cells with a big and obvious nucleus were formed in the primordia (Figure 5H and 5M). However, in 46.7% (28/60) ovules from 35S-TCP3 lines and 42.2% (65/154) ovules from 35S-TCP5 lines, no megasporocytes or megaspores were differentiated (Figure 5I and 5N). In the ovules without megasporocytes, no egg cells, central cells, synergid cells and antipodal cells were formed (Figure 5J, 5K and 5O), resembling the *spl* ovules. The integuments of ovules from 35S-TCP3 plants also exhibited retarded growth (Figure 5J and 5K), a phenotype also observed in *spl/nzz* mutants[6]. Our results suggest that the activities of CIN-like TCPs are essential for normal ovule development and that SPL is an important negative regulator that modulates the activities of TCPs.

Taken together, our results indicate that SPL recruits TPL/TPR co-repressors to suppress the activities of CIN-like TCPs and possibly controls the expression of the TCP downstream genes[27], thus promoting cell differentiation of megasporocytes during ovule development (Figure 5P). When SPL is inactivated, TPL/TPR co-repressors are disassociated from CIN-like TCPs. The excessive activities of TCPs alter the expression of TCP downstream genes and cause the failure of megasporogenesis and abnormal ovule development (Figure 5Q).

DISCUSSION

In this study, we demonstrate that SPL/NZZ controls ovule development by recruiting the TPL/TPR transcription co-repressors to suppress CIN-like TCP transcription factors. First, we show that the five amino acid residues at the C-terminal end of SPL is a typical EAR motif that is required for transcriptional repression activity. Second, SPL physically interacts with TPL/TPR co-repressors through the EAR motif. Third, disruption of TPL/TPRs phenocopies the ovule defects observed in *spl* mutants[5,6]. Fourth, we show that six CIN-like *TCPs* are expressed in ovules and that they interact with SPL. Fifth, disruption of CIN-like TCPs leads to abnormal ovule arrangement, a phenotype also observed in *spl-D*gain-of-function mutants. Finally, either overexpression of the CIN-like *TCPs* or inactivation of *SPL* results in aborted ovules. These findings reveal the SPL action mechanism: SPL functions as an adaptor-like transcriptional repressor linking TPL/TPR co-repressors to the key transcription factors in the megasporogenesis during ovule development.

SPL is also known to regulate pollen development in *Arabidopsis*[5,9]. It is not clear whether SPL regulates pollen development by a similar mechanism, i.e., by recruiting TPL/TPR co-repressors to certain transcription factors. However, we did find that *TPL* was highly expressed in stamens and pollens, and that *tpl-1* mutants also produced defective pollens. The fact that pSPL-SPLΔEAR-TPLC complemented not only the female sterility but also the male sterility of *spl-2*strongly suggests that SPL-TPL/TPRs may also function in pollen development Recently, TPL/TPR co-repressors were reported to directly interact with EAR motif-containing transcription repressors DAZ1 and DAZ2 in microgametogenesis and male germ cell division[29], supporting that TPL/TPRs-mediated gene repression plays pivotal roles not only in ovule development but also in pollen development.

The action mechanism of SPL discovered in this study is analogous to those employed in auxin signaling[17], jasmonic acid (JA) signaling[18], circadian clock regulation[30], meristem maintenance[31,32], floral organ development[21] and leaf development[19]. The common feature of these pathways is that the TPL/TPR co-repressors are recruited, through an adaptor-like repressor or not, to repress the activities of certain transcription factors. For example, TPL/TPR co-repressors are directly recruited to transcription factors that contain the EAR motifs, i.e., PRRs, WUSCHEL (WUS), RAMOSA1 (RA1) and APETALA2 (AP2), to regulate circadian clock, meristem maintenance, and floral organ development, respectively[21,30,32,33]. In auxin signaling, JA signaling and leaf development, however, the adaptor-like repressors (i.e., AUXs/IAAs, NINJA, and TIEs, respectively), through their EAR motifs, recruit TPL/TPR co-repressors to repress AUXIN RESPONSIVE FACTORs, MYCs, and TCP transcription factors, respectively[17,18,19]. In this study, we found that, in ovule development, the adaptor-like repressor SPL recruited TPL/TPR co-repressors through its C-terminal EAR motif and interacted with CIN-like TCP transcription factors through its N-terminal part without the EAR motif, suggesting that similar mechanism has been adopted to regulate ovule development as in auxin and JA signaling, and leaf development regulation. The ovules are the main organs for plant reproduction, whereas leaves are the main organs for vegetative growth. Interestingly, both SPL and TIEs recruit TPL/TPR co-repressors and regulate CIN-like TCP activities in ovules and leaves. This supports the hypothesis that similar regulatory mechanisms are adopted in ovules and leaves during plant evolution[19,34]. However, unlike TIEs-TPL/TPR complexes, which delay leaf cell differentiation[19], SPL-TPL/TPR complexes actually promote the differentiation of the archesporial cell into the megasporocytes and promote the growth of ovule integuments.

An ovule can be distinguished as the proximal funiculus which is the stalk linking an ovule to the ovary wall, the central integuments and the distal nucellus[7]. SPL participates in the formation of all the three parts[7,8,12]. SPL controls the cell number of the proximal funiculus by repressing the expression of *ANT* and *INO*[7]. In the central region, SPL and BEL1 coordinate to determine the identity of ovule chalaza where the integuments and nucellus are joined, since the chalaza of *nzz bel1* double mutant was partially substituted by the tissue of the funiculus[7]. SPL also interacts with *INO* and represses the expression of *INO* and *ANT* to control outer integument development and thus the formation of ovule adaxial-abaxial axis[8,12]. In the distal region, SPL represses the expression of *ANT, BEL1*, and *INO*in the formation of the nucellus and/or the megasporocyte[7]. These findings suggest that SPL acts as a repressor in these ovule developmental processes; but how SPL suppress these genes at the molecular level has long been unclear. Based on the working model of SPL from this study, it is possible that SPL connects CIN-like TCPs and/or other transcriptional factors with the TPL/TPR co-repressors and thus represses the expression of *ANT, BEL1*, and *INO*. However, it seems that the roles of SPL are rather complicated during ovule development. In addition to repressing the transcription of *ANT, BEL1*, and *INO*, SPL also promotes the expression of *PIN1* and *WUS* in the ovule[35]. The possible explanation might be that SPL represses the activities of some unknown suppressors and releases the transcription of *PIN1* and/or *WUS*. The finding that SPL interacts with TPL/TPRs and TCPs during ovule development adds a new layer of the complexity of the SPL-mediated regulation network. Whether *ANT, BELL1*, and *INO* are direct target genes of CIN-like TCPs and whether TCPs and INO bind to similar regions of SPL need to be investigated in the future.

TCP transcription factors are grouped into two classes, Class I TCPs and Class II TCPs, based on the sequence similarities of the TCP domains[36]. CINCINNATA (CIN)-like TCPs belong to the Class II TCPs and promote leaf cell differentiation[36]. CIN-like TCPs have long been reported to play pivotal roles in leaf development, because disruption of CIN-like TCPs causes excessive growth at the leaf margins in *Antirrhinum* and *Arabidopsis*[24,37]. In addition to the CIN-like *TCP* genes, most of the Class I *TCP* genes are also expressed in leaves[26,38,39], but Class I TCPs are functionally redundant, because even the pentuple mutants of the Class I *TCP*genes (i.e., *tcp8 tcp15 tcp21 tcp22 tcp23*) have almost no obvious phenotypes[38]. Because the Class I TCPs and Class II TCPs bind to different optimal *cis*-regulatory elements but with overlapping consensus sequences, it is possible that two classes of TCPs compete with the same binding sites[36]. It is clear that CIN-like TCPs play essential roles in ovule development, but whether Class I TCPs are also involved in ovule development still needs verification.

SPL controls multiple layers of plant development by affecting cell fate determination, cell differentiation, and organ identity[5,6,7,8,9,10,11]. This work not only demonstrates the essential roles of CIN-like TCP transcription factors and TPL/TPR co-repressors in ovule development, but also clarifies the mechanism of how these two groups of proteins are involved in the SPL-mediated regulation of ovule development. The temporal and spatial TCP activities are essential for organ development[26,40]. The recruitment of TPL/TPR co-repressors by SPL to the TCPs revealed an accurate and flexible mechanism in control of the CIN-like TCP activities in the megasporogenesis during ovule development. Interestingly, disruption of Tup1, the TPL ortholog in *S. cerevisiae*, or disruption of TPL orthologs in fungal pathogens including *Penicillium marneffei, Candida albicans,Neurospora crassa* and *Ustilago maydis*, led to defective spore production[41,42,43,44,45,46], suggesting that Tup1 co-repressors are essential for sporogenesis in fungi, a process similar to plant sporogenesis. Notably, Tup1 needs to be recruited to specific transcription factors by adaptor protein Suppressor of snf1 6 (Ssn6) and the homozygous *ssn6* diploid *S. cerevisiae* fails to sporulate[47]. Our findings suggest that TPL/Tup1 co-repressor-involved mechanism probably represents an evolutionarily conserved strategy to control sporogenesis by both plants and fungi.

MATERIALS AND METHODS

Plant Materials and Growth Conditions

The *Arabidopsis thaliana* Columbia-0 (Col-0) ecotype and Landsberg *erect* (Ler) ecotype for *topless-1* (*tpl-1*) were used. The *spl-D* was a gain-of-function mutant isolated from our T-DNA insertion mutant collection[11]. The *spl-2* (*SAIL_519_H07*) was obtained from Syngenta *Arabidopsis* Insertion Library (SAIL) collection. The seeds of the dominant-negative allele *tpl-1* were provided by Dr Jeff A Long (University of California, Los Angeles). Dr Tomotsugu Koyama (Kyoto University) kindly provided the seeds of *tcp3 tcp4 tcp5 tcp10* multiple knock-out mutants[27]. The *Arabidopsis* seeds were grown on half-strength Murashige and Skoog (1/2 MS) medium. The herbicide DL-phosphinothricin (20 μg/ml), kanamycin (50 g/ml) or hygromycin (50 μg/ml) was supplemented in 1/2 MS medium for screening the mutants or transgenic plants. The *Arabidopsis* seeds were synchronized in 4 °C refrigerator for 3 days and then were grown in the growth chamber with the temperature of 22 ± 2 °C and long day condition (16-h light and 8-h dark). After 7 days, the green seedlings were transferred to soil and grown in the green house under the same conditions as described above. The *Nicotiana benthamiana* was grown in the same green house for SPL/NZZ transcriptional activity assays, the BiFC analysis and firefly luciferase complementation imaging assays. The primers used in this work are provided in Supplementary information, Table S1.

PCR Analysis and Gene Expression Assays

The primers of LB3, spl-F, spl-R were used for genotyping *spl-2* mutant. The genotyping analysis of *spl-D* and *jaw-5D* was carried out as described previously[11,19]. PCR were performed using the cycle condition: 94 °C for 30 s, 56 °C for 30 s, and 72 °C for 1 min.

For quantitative RT-PCR, total RNAs of the seedlings, flowers, ovules or pistils from WT, *spl-D, tpl-1, tpl-1/spl-D* or *jaw-5D* were extracted using TRIzol reagent (Invitrogen). Total RNAs were reverse-transcribed and quantitative RT-PCR were performed as described previously[48,49]. Briefly, three biological repeats were executed using a SYBR Green real-time PCR Master Mix (TOYOBO) using ABI 7500 Fast Real-Time PCR System (Applied Biosystems). The $2^{-\Delta\Delta CT}$ method was used to calculate the relative expression level of each gene[50]. *ACT8* was used as an internal control.

Generation of Binary Constructs and Transformation

To generate pSPL-SPLΔEAR-TPLC, the 4.4-kb-long *SPL* promoter was amplified from genomic DNA of WT *Arabidopsis* using the primer pair of pATSP-attB4 and pATSP-attBr1 or that of pSPL-F and pSPL-R. The fragment was cloned into pDONRP4P1R (Invitrogen) to generate pEN-L4-4.4k-R1 by the BP reaction or into pENTR/D-TOPO to get pENTRY-4.4k by TOPO cloning. PSPL-GUS was generated by LR reaction between pENTRY-4.4k and pKGWFS7. SPL-F and SPL-TPLC-EAR-R primer pair was used for amplifying the fragment containing SPLΔEAR with a linker of a short 5′-end of TPLC. TPLC-SPL-EAR-F and TPL-R primer pair was used for amplifying the fragment containing a linker of 3′-end of SPLΔEAR with TPLC. The two fragments were denatured and annealed to be template for amplifying the SPLΔEAR-TPLC fusion with the primers SPL-F and TPL-R. The SPLΔEAR-TPLC fusion fragment was cloned into pENTR/D-TOPO to generate pENTRY-SPLΔEAR-TPLC. pSPL-SPLΔEAR-TPLC was generated by LR reaction with the plasmids of pEN-L4-4.4k-R1, pH7m24GW3 and pENTRY-SPLΔEAR-TPLC. To generate pSPL-SPLΔEAR, *SPLΔEAR* were amplified from cDNA of *spl-D* with the primer pairs of SPL-F and SPL-ΔEAR-R. The fragment was cloned into pENTR/D-TOPO to generate pENTRY-SPLEAR. pSPL-SPLΔEAR was generated by LR reaction with the plasmids of pEN-L4-4.4k-R1, pH7m24GW3 and pENTRY-SPLΔEAR.

To generate 35S-SPLmEAR construct, *SPLmEAR* which encodes the SPL protein containing a mutated EAR motif was amplified from cDNA of *spl-D* with the primer pair of SPL-F and SPL-mEAR-R. The DNA fragment was cloned into pENTR/D-TOPO to generate pENTRY-

SPLmEAR. 35S-SPLmEAR was generated by LR reaction with the plasmids of pENTRY-SPLmEAR and pK2GW7.

To generate 35S-TCP17-SPLC, the sequence encoding C-terminus of SPL (from residue 157 to 314) was amplified using primers SPLC-F and SPLC-R and cloned into pDONRP2rP3 (Invitrogen) to generate pEN-R2-SPLC-L3. The pENTRY-TCP17N and pEN-L4-35S-R1 were generated as described previously. 35S-TCP17-SPLC was generated by LR reactions using the plasmids of pK7m34GW, pEN-L4-35S-R1, pENTRY-TCP17N and pEN-R2-SPLC-L3. To generate 35S-TCP17 and 35S-SPLC, the coding region of TCP17 or the C-terminus of SPL was amplified using primers TCP17-1 and TCP17-2 or SPLC-1 and SPLC-2. The fragments were cloned into pENTR/D-TOPO to generate pENTRY-TCP17 or pENTRY-SPLC. 35S-TCP17 or 35S-SPLC was generated by LR reaction between the plasmids pK2GW7 and pENTRY-TCP17 or pENTRY-SPLC.

To generate TCP overexpression constructs, the coding region of TCP3 or TCP5 was amplified from genomic DNA from *Arabidopsis* by primer pairs of TCP3-1 and TCP3-2 or TCP5-1 and TCP5-2. The fragments were cloned into pENTR/D-TOPO to generate pENTRY-TCP3 or pENTRY-TCP5. 35S-TCP3 or 35S-TCP5 was generated by LR reaction between the plasmids pK2GW7 and pENTRY-TCP3 or pENTRY-TCP5.

Constructs were transformed into *Agrobacterium tumefaciens* GV3101/pMP90 and then into *Arabidopsis* as described previously[51].

Staining and Microscopy

For GUS staining, flowers from pSPL-GUS or TPLP-GUS transgenic lines were soaked in 90% acetone solution for 20 min on the ice. The tissues were washed twice in the phosphate buffer and then were put into GUS staining buffer containing 0.5 mg/ml 5-bromo-4-chloro-3-indolyl glucuronide. The samples were vacuumed for 10 min and were stained overnight in 37 °C incubator. The staining buffer was then changed to 70% ethanol for clearing chlorophyll. The ovules were dissected and observed using microscope (OLYMPUS SPE microscope). The Alexander›s staining of pollens was performed as described previously[49]. The DAPI (4, 6-diamidino-2-phenylindole) staining and the scanning electron microscopy was performed as described previously[19].

The observation of ovules using CLSM was described previously[52]. Briefly, the inflorescences with open flowers removed was fixed in 4% glutaraldehyde (in 12.5 mM cacodylate, pH 6.9), and vacuumed until the tissues were all sunk in the bottom of the container. The inflorescences were fixed overnight at room temperature and then were gradually dehydrated using 15%, 30%, 50%, 70%, 80%, 90%, 95% and 100% ethanol. The tissues were cleared in the solution (2 volume of benzyl benzoate : 1 volume benzyl alcohol) for at least 2 h. The pistils were dissected and the ovules were observed using a confocal laser scanning microscope (Leica TCS SPE confocal microscope).

Yeast Two Hybrid Assays

To test the interaction of SPL and TPL/TPRs, *SPL* was amplified from cDNA of *spl-D*with the primer pairs of SPL-F and SPL-R. The fragment was cloned into pENTR/D-TOPO to generate pENTRY-SPL. The prey constructs of AD-SPL or AD-SPLmEAR were generated by LR reactions between pDEST22 (Invitrogen) and pENTRY-SPL or pENTRY-SPLmEAR. The bait construct of pDEST32-NTPL (DBD-NSPL) was generated as described previously[19]. To confirm the interaction of SPL and TPL/TPRs, we also generated the constructs of DBD-SPL, DBD-SPLmEAR and DBD-SPLΔEAR using LR reactions between pDEST32 (Invitrogen) and pENTRY-SPL, pENTRY-SPLmEAR or pENTRY-SPLΔEAR. The pENTRY-NTPL and pENTRY-NTPRs were generated as described previously[19]. The AD-NTPL and AD-NTPRs were generated by LR reactions between pDEST22 and pENTRY-NTPL or pENTRY-NTPRs. The bait and prey plasmids or the blank pDEST22 were co-transformed into AH109 (Clontech) yeast strain.

To investigate the interactions between CIN-like TCPs and SPL or SPLΔEAR, *TCP*genes were cloned into pDEST22 as preys[19]. Bait plasmids of DBD-SPL or DBD-SPLEAR and prey plasmids of pDEST22-TCPs or the blank pDEST22 were co-transformed into AH109 yeast cells.

The yeast cells were selected on the medium supplemented with 2.5 mM 3-amino-1,2,4 triazole and SD-Leu-Trp-His or SD-Ade-Leu-Trp-His.

Transient Expression, BiFC and Firefly Luciferase Complementation Imaging Assays

The G4BD-SPL, G4BD-SPLmEAR or G4DBD-SPLΔEAR was amplified from DBD-SPL, DBD-SPLmEAR or DBD-SPLΔEAR using primers of G4BD-1 and SPL-R, SPLmEAR-R or SPLΔEAR-R. The PCR products were cloned into pENTR/D-TOPO to generate pENTRY-G4BD-SPL, pENTRY-G4BD-SPLmEAR or pENTRY-G4BD-SPLΔEAR. The effector constructs were obtained by LR reactions between pK2GW7 and pENTRY-G4BD-SPL, pENTRY-G4BD-SPLmEAR or pENTRY-G4BD-SPLΔEAR. The plasmids of effectors were co-transformed into the leaves of *N. benthamiana* with the reporter 35S-UAS-GUS and pCam-P19 using the method mediated by*Agrobacterium* GV3101/pMP90. GUS staining was described previously[19].

To test the interaction between SPL and TPL, the construct cCFP-TPL was generated by LR reaction between pcCFPxGW and pENTRY-TPL[19]. LR reaction was performed to generate nYFP-SPL using the plasmids of pnYFPxGW and pENTRY-SPL. BiFC analysis was performed as described previously with slight modifications. These plasmids were transformed into *Agrobacterium* GV3101/pMP90. The nYFP-SPL was co-transformed with pCam-P19 and cCFP-TPL into the leaves of *N. benthamiana*. The empty vectors of pcCFPxGW and pnYFPxGW were co-transformed with nYFP-SPL and cCFP-TPL as controls. The plants were incubated in dark for 24 h and then in light for 72 h. The leaves were observed using the microscope (Leica TCS SPE confocal microscope).

To test the interactions of SPL and TCPs *in vivo*, SPL was amplified from pENTRY-SPL using SPL-KPN1-F and SPL-SAL1-R. The PCR fragment was digested by *Kpn* I and *Sal* I and cloned into the vector pCAMBIA-nLuc to generate SPL-nLuc[53]. The*TCP* genes were amplified using corresponding primer pairs and cloned into pCAMBIA-cLuc to generate cLuc-TCPx[53]. These plasmids were transformed into*Agrobacterium* GV3101/pMP90. Firefly luciferase complementation imaging assays were carried out as described previously[53]. Briefly, SPL-nLuc and Luc-TCPx, and the control combinations were co-transformed with pCam-P19 into the leaves of *N. benthamiana*, respectively. Every combination was repeated more than three times in different leaves. The plants were incubated in dark for 24 h and then in light for 48 h.

The leaves were sprayed with 1 mM luciferin and observed under a low-light cooled CCD imaging apparatus (lumazone 1300B, Bio-One Scientific Instrument).

ACKNOWLEDGEMENTS

We thank Dr Yunde Zhao (University of California at San Diego, USA) for valuable suggestions. We thank Dr Jeff Long (University of California, Los Angeles, USA) and Dr Tomotsugu Koyama (Kyoto University, Japan) for kindly providing the seeds of *tpl-1* and *tcp3 tcp4 tcp5 tcp10*. This work was supported by the National Key Basic Research Program of China (973-2012CB944801), the National Natural Science Foundation of China (31270321), and National Transformation Science and Technology Program (2014ZX08009003-003-003). This work was also partially supported by the 111 Project (B06001).

REFERENCES

1. Shi DQ, Yang WC. Ovule development in *Arabidopsis*: progress and challenge. *Curr Opin Plant Biol* 2011; 14:74–80.
2. Yang WC, Shi DQ, Chen YH. Female gametophyte development in flowering plants. *Annu Rev Plant Biol* 2010; 61:89–108.
3. Sundaresan V, Alandete-Saez M. Pattern formation in miniature: the female gametophyte of flowering plants. *Development* 2010; 137:179–189.
4. Liu J, Qu LJ. Meiotic and mitotic cell cycle mutants involved in gametophyte development in *Arabidopsis*. *Mol Plant* 2008; 1:564–574.
5. Yang WC, Ye D, Xu J, Sundaresan V. The *SPOROCYTELESS* gene of*Arabidopsis* is required for initiation of sporogenesis and encodes a novel nuclear protein. *Genes Dev* 1999; 13:2108–2117
6. Schiefthaler U, Balasubramanian S, Sieber P, Chevalier D, Wisman E, Schneitz K. Molecular analysis of *NOZZLE*, a gene involved in pattern formation and early sporogenesis during sex organ development in *Arabidopsis thaliana*. *Proc Natl Acad Sci USA* 1999; 96:11664–11669

7. Balasubramanian S, Schneitz K. *NOZZLE* regulates proximal-distal pattern formation, cell proliferation and early sporogenesis during ovule development in *Arabidopsis* thaliana. *Development* 2000; 127:4227–4238.
8. Balasubramanian S, Schneitz K. *NOZZLE* links proximal-distal and adaxial-abaxial pattern formation during ovule development in *Arabidopsis* thaliana.*Development* 2002; 129:4291–4300.
9. Ito T, Wellmer F, Yu H, *et al.* The homeotic protein AGAMOUS controls microsporogenesis by regulation of *SPOROCYTELESS. Nature* 2004;430:356–360.
10. Liu X, Huang J, Parameswaran S, *et al.* The *SPOROCYTELESS/NOZZLE* gene is involved in controlling stamen identity in *Arabidopsis. Plant Physiol* 2009;151:1401–1411.
11. Li LC, Qin GJ, Tsuge T, *et al.* SPOROCYTELESS modulates YUCCA expression to regulate the development of lateral organs in *Arabidopsis. New Phytol*2008; 179:751–764.
12. Sieber P, Petrascheck M, Barberis A, Schneitz K. Organ polarity in*Arabidopsis.* NOZZLE physically interacts with members of the YABBY family.*Plant Physiol* 2004; 135:2172–2185.
13. Cheng Y, Dai X, Zhao Y. Auxin biosynthesis by the YUCCA flavin monooxygenases controls the formation of floral organs and vascular tissues in *Arabidopsis. Genes Dev* 2006; 20:1790–1799.
14. Bencivenga S, Simonini S, Benkova E, Colombo L. The transcription factors BEL1 and SPL are required for cytokinin and auxin signaling during ovule development in *Arabidopsis. Plant Cell* 2012; 24:2886–2897.
15. Ohta M, Matsui K, Hiratsu K, Shinshi H, Ohme-Takagi M. Repression domains of class II ERF transcriptional repressors share an essential motif for active repression. *Plant Cell* 2001; 13:1959–1968.
16. Hiratsu K, Mitsuda N, Matsui K, Ohme-Takagi M. Identification of the minimal repression domain of SUPERMAN shows that the DLELRL hexapeptide is both necessary and sufficient for repression of transcription in *Arabidopsis.Biochem Biophys Res Commun* 2004; 321:172–178.
17. Szemenyei H, Hannon M, Long JA. TOPLESS mediates auxin-dependent transcriptional repression during *Arabidopsis* embryogenesis. *Science* 2008;319:1384–1386

18. Pauwels L, Barbero GF, Geerinck J, *et al*. NINJA connects the co-repressor TOPLESS to jasmonate signalling. *Nature* 2010; 464:788–791.
19. Tao Q, Guo D, Wei B, *et al*. The TIE1 transcriptional repressor links TCP transcription factors with TOPLESS/TOPLESS-RELATED corepressors and modulates leaf development in *Arabidopsis*. *Plant Cell* 2013; 25:421–437.
20. Causier B, Ashworth M, Guo W, Davies B. The TOPLESS interactome: a framework for gene repression in *Arabidopsis*. *Plant Physiol* 2012; 158:423–438
21. Krogan NT, Hogan K, Long JA. APETALA2 negatively regulates multiple floral organ identity genes in *Arabidopsis* by recruiting the co-repressor TOPLESS and the histone deacetylase HDA19. *Development* 2012; 139:4180–4190
22. Long JA, Woody S, Poethig S, Meyerowitz EM, Barton MK. Transformation of shoots into roots in *Arabidopsis* embryos mutant at the TOPLESS locus.*Development* 2002; 129:2797–2806.
23. Ou B, Yin KQ, Liu SN, *et al*. A high-throughput screening system for*Arabidopsis* transcription factors and its application to Med25-dependent transcriptional regulation. *Mol Plant* 2011; 4:546–555.
24. Palatnik JF, Allen E, Wu X, *et al*. Control of leaf morphogenesis by microRNAs. *Nature* 2003; 425:257–263.
25. Schommer C, Palatnik JF, Aggarwal P, *et al*. Control of jasmonate biosynthesis and senescence by miR319 targets. *PLoS Biol* 2008; 6: e230.|
26. Koyama T, Furutani M, Tasaka M, Ohme-Takagi M. TCP transcription factors control the morphology of shoot lateral organs via negative regulation of the expression of boundary-specific genes in *Arabidopsis*. *Plant Cell* 2007;19:473–484.
27. Koyama T, Mitsuda N, Seki M, Shinozaki K, Ohme-Takagi M. TCP transcription factors regulate the activities of ASYMMETRIC LEAVES1 and miR164, as well as the auxin response, during differentiation of leaves in *Arabidopsis*. *Plant Cell* 2010; 22:3574–3588.
28. Alvarez-Buylla ER, Benitez M, Corvera-Poire A, *et al*. Flower development.*Arabidopsis Book* 2010; 8: e0127.

29. Borg M, Rutley N, Kagale S, *et al*. An EAR-dependent regulatory module promotes male germ cell division and sperm fertility in *Arabidopsis*. *Plant Cell*2014; 26:2098–2113.

30. Wang L, Kim J, Somers DE. Transcriptional corepressor TOPLESS complexes with pseudoresponse regulator proteins and histone deacetylases to regulate circadian transcription. *Proc Natl Acad Sci USA* 2013; 110:761–766.|

31. Kieffer M, Stern Y, Cook H, *et al*. Analysis of the transcription factor WUSCHEL and its functional homologue in *Antirrhinum* reveals a potential mechanism for their roles in meristem maintenance. *Plant Cell* 2006; 18:560–573.

32. Gallavotti A, Long JA, Stanfield S, *et al*. The control of axillary meristem fate in the maize *ramosa* pathway. *Development* 2010; 137:2849–2856.

33. Ikeda M, Mitsuda N, Ohme-Takagi M. *Arabidopsis* WUSCHEL is a bifunctional transcription factor that acts as a repressor in stem cell regulation and as an activator in floral patterning. *Plant Cell* 2009; 21:3493–3505.

34. Kelley DR, Gasser CS. Ovule development: genetic trends and evolutionary considerations. *Sex Plant Reprod* 2009; 22:229–234.

35. Sieber P, Gheyselinck J, Gross-Hardt R, Laux T, Grossniklaus U, Schneitz K. Pattern formation during early ovule development in *Arabidopsis* thaliana.*Dev Biol* 2004; 273:321–334.

36. Martin-Trillo M, Cubas P. TCP genes: a family snapshot ten years later.*Trends Plant Sci* 2010; 15:31–39.

37. Nath U, Crawford BC, Carpenter R, Coen E. Genetic control of surface curvature. *Science* 2003; 299:1404–1407.

38. Aguilar-Martinez JA, Sinha N. Analysis of the role of *Arabidopsis* class I *TCP*genes *AtTCP7*, *AtTCP8*, *AtTCP22*, and *AtTCP23* in leaf development. *Front Plant Sci* 2013; 4:406.

39. Herve C, Dabos P, Bardet C, *et al*. *In vivo* interference with AtTCP20 function induces severe plant growth alterations and deregulates the expression of many genes important for development. *Plant Physiol* 2009;149:1462–1477.

40. Efroni I, Blum E, Goldshmidt A, Eshed Y. A protracted and dynamic maturation schedule underlies *Arabidopsis* leaf development. *Plant Cell* 2008;20:2293–2306.

41. Elias-Villalobos A, Fernandez-Alvarez A, Ibeas JI. The general transcriptional repressor Tup1 is required for dimorphism and virulence in a fungal plant pathogen. *PLoS Pathog* 2011; 7: e1002235

42. Friesen H, Hepworth SR, Segall J. An Ssn6-Tup1-dependent negative regulatory element controls sporulation-specific expression of *DIT1* and *DIT2*in *Saccharomyces cerevisiae. Mol Cell Biol* 1997; 17:123–134.

43. Todd RB, Greenhalgh JR, Hynes MJ, Andrianopoulos A. TupA, the *Penicillium marneffei* Tup1p homologue, represses both yeast and spore development.*Mol Microbiol* 2003; 48:85–94.

44. Braun BR, Johnson AD. Control of filament formation in *Candida albicans* by the transcriptional repressor TUP1. *Science* 1997; 277:105–109.

45. Yamashiro CT, Ebbole DJ, Lee BU, *et al*. Characterization of *rco-1* of*Neurospora crassa*, a pleiotropic gene affecting growth and development that encodes a homolog of Tup1 of *Saccharomyces cerevisiae. Mol Cell Biol*1996; 16:6218–6228.

46. Mizuno T, Nakazawa N, Remgsamrarn P, Kunoh T, Oshima Y, Harashima S. The Tup1-Ssn6 general repressor is involved in repression of IME1 encoding a transcriptional activator of meiosis in *Saccharomyces cerevisiae. Curr Genet* 1998; 33:239–247.

47. Schultz J, Carlson M. Molecular analysis of *SSN6*, a gene functionally related to the *SNF1* protein kinase of *Saccharomyces cerevisiae. Mol Cell Biol* 1987;7:3637–3645.

48. Zhao Y, Wei T, Yin KQ, *et al. Arabidopsis* RAP2.2 plays an important role in plant resistance to *Botrytis cinerea* and ethylene responses. *New Phytol*2012; 195:450–460.

49. Wang WY, Zhang L, Xing S, *et al. Arabidopsis* AtVPS15 plays essential roles in pollen germination possibly by interacting with AtVPS34. *J Genet Genom*2012; 39:81–92

50. Livak KJ, Schmittgen TD. Analysis of relative gene expression data using real-time quantitative PCR and the 2(-Delta Delta C(T)) method. *Methods*2001; 25:402–408.

51. Qin G, Gu H, Zhao Y, *et al*. An indole-3-acetic acid carboxyl methyltransferase regulates *Arabidopsis* leaf development. *Plant Cell* 2005;17:2693–2704.

52. Shi DQ, Liu J, Xiang YH, Ye D, Sundaresan V, Yang WC. *SLOW WALKER1*, essential for gametogenesis in *Arabidopsis*, encodes a WD40 protein involved in 18S ribosomal RNA biogenesis. *Plant Cell* 2005; 17:2340–2354.

53. Chen H, Zou Y, Shang Y, *et al*. Firefly luciferase complementation imaging assay for protein-protein interactions in plants. *Plant Physiol* 2008;146:368–376.

Chapter 2

Parasitic Plants have Increased Rates of Molecular Evolution across all Three Genomes

Lindell Bromham, Peter F Cowman, and Robert Lanfear

Centre for Macroevolution, and Macroecology, Research School of Biology, Australian National University, Canberra, A.C.T, 0200, Australia

ABSTRACT

Background

Theoretical models and experimental evidence suggest that rates of molecular evolution could be raised in parasitic organisms compared to non-parasitic taxa. Parasitic plants provide an ideal test for these predictions, as there are at least a dozen independent origins of the parasitic lifestyle in angiosperms. Studies of a number of parasitic plant lineages have suggested faster rates of molecular evolution, but the results of some studies have been mixed. Comparative analysis of all parasitic plant lineages, including sequences from all three genomes, is needed to examine the generality of the relationship between rates of molecular evolution and parasitism in plants.

Results

We analysed DNA sequence data from the mitochondrial, nuclear and chloroplast genomes for 12 independent evolutionary origins of parasitism in angiosperms. We demonstrated that parasitic lineages have a faster rate of molecular evolution than their non-parasitic relatives in sequences for all three genomes, for both synonymous and no synonymous substitutions.

Conclusions

Our results prove that raised rates of molecular evolution are a general feature of parasitic plants, not confined to a few taxa or specific genes. We discuss possible causes for this relationship, including increased positive selection associated with host-parasite arms races, relaxed selection, reduced population size or repeated bottlenecks, increased mutation rates, and indirect causal links with generation time and body size. We find no evidence that faster rates are due to smaller effective population's sizes or changes in selection pressure. Instead, our results suggest that parasitic plants have a higher mutation rate than their close non-parasitic relatives. This may be due to a direct connection, where some aspect of the parasitic lifestyle drives the evolution of raised mutation rates. Alternatively, this pattern may be driven by an indirect connection between rates and parasitism: for example, parasitic plants tend to be smaller than their non-parasitic relatives, which may result in more cell generations per year, thus a higher rate of mutations arising from DNA copy errors per unit time. Demonstration that adoption of a parasitic lifestyle influences the rate of genomic evolution is relevant to attempts to infer molecular phylogenies of parasitic plants and to estimate their evolutionary divergence times using sequence data.

BACKGROUND

Theoretical models have led to the prediction that parasites should evolve higher mutation rates in order to out-evolve their hosts [1]. This prediction has received support from experimental populations of bacteria [2, 3]. But does it also apply to multicellular eukaryotes? A number of molecular phylogenetic studies have noted that parasitic taxa have much longer branch lengths, suggesting elevated rates of molecular evolution [4-7]. While it would not be surprising to find that specific genes involved in host-parasite interaction experience elevated substitution rates because of specific selection pressures [8], these studies report higher substitution rates in "housekeeping" genes that are not specifically connected to host-parasite interaction. It is therefore pertinent to ask whether being a parasite causes an increase in the genome-wide rate of molecular evolution in multicellular organisms.

Parasitic plants represent an ideal case for testing the hypothesis that parasites have faster rates of molecular evolution. They show a range of adaptations to parasitism and vary in the type and degree of interaction with their hosts [9, 10]. Non-photosynthetic holoparasites are completely dependent on their host for energy, whereas hemiparasites retain photosynthetic ability. Hemiparasites can be either obligate (host-dependent) or facultative (able to be free-living). Parasitic plants also differ in their physical connection to their host (e.g., through the stems or roots), and the degree of reduction of their own anatomy (e.g., loss of stems, leaves or roots).

Faster substitution rates have been reported for a number of genes in several parasitic plant lineages [6, 7, 11-14]. Furthermore, the mode of parasitism has been suggested to influence rates of molecular evolution, for example holoparasites have been shown to have faster rates than confamilial hemiparasites in three plastid genes of the diverse broomrape family Orobanchaceae[4]. However, not all studies have demonstrated a general increase in substitution rate in parasitic plants. For example, a study of substitution rates in the 18S rRNA gene found that less than half of the parasitic taxa had elevated rates of molecular evolution (3 out of 16 hemiparasitic taxa, and 8 out of 17 holoparasites [7]), and only some of the nuclear genes in a study of substitution rates in the holoparasitic root parasite *Balanophora* showed significantly increased rates [13]. So the generality and possible causes of the increase in rates in parasite genomes remain unknown.

There are several reasons why a wide-ranging comparative study of rates of molecular evolution in parasitic plants is needed. Most importantly, a comparative study allows us to move from anecdotal observation to general principle. Although faster rates of molecular evolution have been noted in several different species of parasitic plants, we cannot tell from those studies alone whether accelerated rates are a general consequence of their parasitic lifestyle, or whether the faster rates in these particular species and genes are incidental to parasitism. By examining rate variation for all available parasitic lineages in the same analysis we can test the generality of the pattern.

A phylogenetic comparative study also provides a statistically rigorous way to test the hypothesis that parasitic lifestyle increases the rate of molecular evolution. Previously published studies have used a range of data, different methods, and different statistical approaches. In particular, many have used overlapping comparisons between parasites and non-parasites [6, 13, 15], or treated each parasitic lineage as an independent data point [7], violating assumptions of statistical independence inherent in the tests [16]. It is therefore important to analyse all of the available data within a single statistical framework to describe general patterns in molecular evolution in all independent lineages of parasitic plants.

A comparative study also allows us to compare all three plant genomes, to check for consistency of patterns of molecular evolution. For example, we might expect that the chloroplast genome would evolve rapidly if released from selective constraint, but it is not clear that the same explanation would apply to genes from the mitochondrial or nuclear genomes [7]. Studies that have reported higher rates in parasitic plants have not tested different possible explanations of this pattern, such as positive selection, relaxed purifying selection, or changes in population size. By comparing different substitutions rates within the same analytical framework, we can examine the reasons for the proposed acceleration in rates in parasitic plant taxa, by teasing apart the patterns due to mutation rate, selection and population processes.

There are many independent origins of parasitic lifestyle in angiosperms [17], so we are able to analyses data from 12 separate lineages of parasites. These lineages vary widely in growth form, life cycle, distribution and level of specialization to parasitism. We are able to combine sequence data from chloroplast, nuclear, and mitochondrial sequences for representative taxa from all 12 lineages with data from their nonparasitic relatives. This provides sufficient data to allow the relationship between parasitism and faster rates to be examined within a single phylogenetic comparative framework.

RESULTS

Parasitic plant clades had significantly higher substitution rates in all measures of branch length for all three genomes (Figure 1). Ten out of twelve nuclear comparisons showed a faster substitution rate in the parasitic lineage compared to its nonparasitic relative (Wilcoxon signed ranks test: $Z=2.786$, $p=0.005$). For the mitochondrial genes, all but one comparison had longer branch lengths in the parasitic clade ($Z=2.628$, $p=0.009$) and both nonsynonymous (dN: $Z=2.942$, $p=0.003$) and synonymous rates ($Z=2.157$, $p=0.031$) were significantly higher in the parasitic lineages.

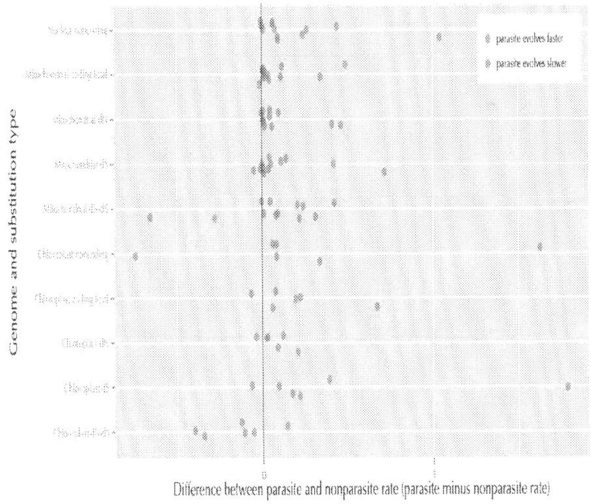

Figure 1: Scatter plot of comparisons between phylogenetically independent pairs of parasitic plants and their nonparasitic relatives.Points were calculated as the value of the nonparasitic clade subtracted from the parasitic clade, so that if the parasite has a faster rate than the difference is greater than 0.

Each line represents a single substitution class in one of the three plant genomes, and each comparison is represented either by a red dot if the parasitic clade has the greater substitution rate or a blue dot if the nonparasitic clade has the higher substitution rate. The points have been slightly jittered to allow overlapping points to be seen clearly.

For the chloroplast sequences, there were fewer sequences available, and neither the rRNA nor the protein-coding sequences had significantly different rates in parasites under the Wilcoxon signed ranks test. However, there was a clear trend in the data. For those comparisons with 16S rRNA (*rrn16*), seven out of eight comparisons has faster rates in the parasitic lineage. For the chloroplast protein-coding alignment, five out of six comparisons had faster rates in the parasitic lineage. We tested for a significant pattern across all comparisons by asking whether the parasite had the faster rate in whichever chloroplast genes were available, considering only the sign but not the magnitude of the branch length difference. Ten out of twelve comparisons had faster rates in the parasitic clade in the chloroplast sequences available for that comparison (two tailed sign test, $p = 0.039$). Because there are only 6 comparisons with chloroplast protein coding genes, it is difficult to draw any conclusions regarding synonymous and nonsynonymous substitution rates, but they do follow the same pattern as the mitochondrial and nuclear genomes, with both raised dS and dN in most parasite lineages, but not increased dN/dS.

We did not find a consistent difference in GC content between sequences from parasitic and non-parasitic taxa for nuclear (signs test: $p = 0.77$), mitochondrial ($p = 0.15$) or chloroplast ($p = 0.15$) alignments. This suggests that GC bias is unlikely to explain the consistent differences in rate between parasites and their non-parasitic relatives.

DISCUSSION

Our analysis provides strong support for the hypothesis that parasitic plants have consistently faster rates of molecular evolution than their non-parasitic relatives in mitochondrial, chloroplast, and nuclear sequences. The pattern is detected in all classes of substitutions tested, including ribosomal rRNA genes, and both synonymous and nonsynonymous substitutions in protein coding genes.

What could increase the rate of substitution so consistently in such varied lineages of parasitic plants? The parasitic plants included in this study range from fungus-like holoparasites, such as the achlorophyllous members of the Apodanthaceae that exist entirely within host tissues except when flowering, to perennial shrubs and trees, such as the sandalwoods (e.g. *Nestronia*) [18]. Here, we will consider several possible explanations for faster substitution rates in parasites.

Relaxed Selection

By drawing on metabolites from another individual, a parasite may be under less pressure to maintain their own means to produce those resources. Indeed, there may be selection pressure to lose unnecessary features, in order to save resources or accelerate the life cycle. Many holoparasites have lost or greatly reduced their leaves and stems, and in extreme cases the main vegetative body of the parasitic plant consists of roots growing as a mycelium-like mass inside the host tissues (e.g. Rafflesiaceae, Cytinaceae, Apodanthaceae). In such circumstances we would expect the genes that make products and structures needed for photosynthesis, such as leaves and chlorophyll, to decay into pseudogenes or be lost from the genome [19-22].

A gene that has no selective pressure to maintain function should acquire substitutions at a rate determined by the mutation rate. In this case, we would expect the nonsynonymous substitution rate to approximate the synonymous rate, and their ratio to be approximately 1. We do not find any evidence to support this effect in our study, where values of the ratio of dN to dS are not consistently closer to one in the parasitic lineages, even in the available protein-coding genes of the chloroplast [14]. It is possible that dN/dS values less than one in the parasitic lineage reflect past (negative) selection. Since our substitution rates are measured from the common ancestor of the sister clades, the genes in the parasitic clade will have been in autotrophic plants for some part of their history. In this case, we might expect dN/dS ratios to be higher (closer to one) in most parasitic plants when compared to their non-parasitic relatives, as they would have accumulated relatively more nonsynonymous substitutions during the parasitic part of their evolutionary history. But there is no clear trend toward higher dN/dS in parasitic plant lineages in any of the sequences: parasitic lineages have higher dN/dS in 9 out of 12 comparisons for mitochondrial genes ($Z = 1.294$, $p = 0.196$), and only 1 out of 6 comparisons for chloroplast genes ($Z = -1.26$, $p = 0.21$).

Loss of function in photosynthesis genes might not follow a simple decay from the adoption of parasitic lifestyle onwards. Chloroplast genes in different species of *Orobanche* (broomrape), a non-photosynthetic root parasite, show variable rates of decay into pseudogenes [23], and parasitic plants show different degrees of loss of genes from the plastid genome [20]. Some parasitic plants retain low levels of photosynthetic activity, insufficient to live independently, but enough to tide them over while waiting to find a suitable host [24]. Adaptation to parasitism may come about through changes to gene expression levels, leaving less of a signature in the coding sequences of genes [25].

Chloroplast genes in parasites are not necessarily divorced from the effects of selection. Chloroplasts genomes are not solely directed toward photosynthesis, they also contain genes that play a variety of key metabolic roles [19, 26]. In any case, it is difficult to imagine that all the genes in this study are under relaxed purifying selection in parasites. The nuclear genes analysed code for essential parts of the protein synthesis machinery, and the mitochondrial genes are associated with gene expression and metabolism. Furthermore, we would not expect relaxation of selection to have a marked effect on synonymous substitution rates. So it does not seem very likely that the acceleration in both synonymous and nonsynonymous substitution rates across all three genomes in a wide range of parasitic plants can be explained through general relaxed selection on gene function.

Positive Selection

Hosts may use a number of different strategies to resist parasitic plants, operating at all stages of parasite establishment, ranging from immune-like responses, expression of parasite-specific toxins, and silencing host genes that make products essential for parasite growth [27]. So parasites will probably have to battle their hosts on multiple fronts, suggesting that many parts of parasite physiology and development, and consequently many different genes, will be under the strong selective pressure of a host-parasite arms-race. Indeed, transcript omics of tissues at the host-parasite interface suggest that many pathways upregulated in the parasites are "general purpose" genes involved in core processes in both parasite and host species [28]. So selection could raise the substitution rate in a wide variety of genes involved in successful exploitation of hosts, including genes that might be considered "housekeeping genes" [5]. The arms race between host and parasite may account for the divergence of parasitic lineages onto different host plant populations, such that parasites can be more successfully grown on their local host plants than those from another population [23].

Positive selection on protein-coding genes could raise the non-synonymous substitution rate, as it could drive the fixation of mutations that altered the functional properties of the protein. It is possible, therefore, that positive selection could explain part of the elevation of non-synonymous substitution rates we observed in parasitic plant lineages. However, we would not expect the synonymous substitution rate to be elevated by a host-parasite arms race of this kind. While some synonymous mutations could experience a degree of selective pressure [29], for example for translational efficiency [30,31], it is difficult to see how this could differ sufficiently between parasitic and autotrophic plants to explain the acceleration in synonymous substitution rates in parasites seen in this study. So if positive selection had raised the non-synonymous substitution rate but left the synonymous substitution rate relatively unchanged in parasitic lineages, then we would expect these lineages to show increased dN/dS relative to non-parasitic lineages. We observe no consistent changes in dN/dS associated with parasitic lineages in our data, suggesting that positive selection is unlikely to explain any significant proportion of the elevated parasitic substitution rates that we have observed.

Reduction in Effective Population Size

Effective population size (N_e) affects patterns and rates of substitution, so should always be considered when seeking an explanation for consistent patterns in molecular evolution [32-34]. Parasitic taxa might undergo more frequent and severe founder effects, which might decrease their long-term effective population size and so increase the rate of substitution [5, 35]. Endosymbiotic bacteria and fungi have higher substitution rates than their free-living relatives, which has been interpreted as a result of consistently lower population sizes and frequent population bottlenecks [36]. Could small population size or frequent population bottlenecks explain the faster rates in parasitic plants?

It is not clear that parasitic plants will typically have significantly smaller effective population sizes than their autotrophic relatives. Many are animal pollinated, indeed the flowers may be the most conspicuous part of the life cycle, like the gigantic fly-pollinated *Rafflesia* flowers, and many species have edible seeds that are likely to be dispersed by mammals or birds [18]. Some parasitic plants are widespread generalists that can tolerate a range of conditions and grow on a variety of host plants, to the extent that they are economically important weeds of many different crops [37]. So there seems no *a priori* reason to suspect that parasitic plants are less likely to be outcrossed or widely dispersed compared to their autotrophic relatives. In any case, lower average population sizes in parasitic lineages should mimic the effects of relaxed purifying selection, on a genome-wide scale, because selection becomes less efficient at removing slightly deleterious mutations in smaller populations [38]. We don't detect any evidence of this in our study, because there is no significant increase in the ratio of nonsynonymous to synonymous substitutions in the parasitic taxa.

Increased Mutation Rate

Parasites will often have much faster rates of molecular evolution than their hosts, potentially due to their much faster rate of generation turnover [39], which may help them get ahead in the arms race [40]. But in this study, we have not compared the rate of molecular evolution of parasites to their hosts, but of parasites to their free-living relatives. Synonymous substitution rate (dS) is commonly interpreted as reflecting the mutation rate, because changes to the gene sequence that do not cause a change in the protein-product are likely to be invisible to selection [41]. However, other factors could influence the synonymous substitution rate, such as GC biased gene conversion [42] and codon-usage bias [29], and synonymous substitution rate can vary across the genome [29, 43]. But while dS is not a perfect predictor of mutation rate, changes in mutation rate should be reflected in changes in the rate of both synonymous and nonsynonymous substitution. So elevated mutation rates in parasites could produce the pattern we observe in this study, raising both the synonymous and non-synonymous substitution rates, but not influencing the dN/dS ratio. If elevated mutation rates can explain the patterns we have observed, we must explain how adopting a parasitic lifestyle could result in an increase in the mutation rate.

Simple models that set up an arms race between host and parasite loci predict that increased mutation rate will allow parasites to "chase" the host's changing defences [1]. Mutator alleles can speed evolution in a clonally reproducing population even when they are not fixed in the population [44]. However, the increase in mutation predicted by chase models is typically transient [45], as the positively selected mutations recombine away from mutator alleles. So alleles that cause a global increase in mutation rate are unlikely to be maintained by selection for the novel traits they generate [46]. The link between mutation rate and adaptive evolution is complex even in asexual lineages [47, 48], and in sexual populations, it is not clear the extent to which adaptation will be mutation limited [49]. It seems unlikely that a sexually reproducing parasitic plant would gain enough advantage from increased mutation rate, in terms of evading host defence, for selection on mutator alleles to outweigh the costs of producing more deleterious mutations.

Alternatively, increased mutation rate in parasites might results from an erosion of the DNA repair systems that maintain replication fidelity and fix incidental damage. For example, it has been suggested that the lower bound of the mutation rate is set by the efficiency of selection for maintenance or improvement to DNA repair, which will be primarily limited by the effective population size [50]. In this case, we might expect to see that dS correlates with N_e, so if parasitic plants had consistently lower population sizes it may increase dS and dN, potentially accounting for the lack of correlation between parasitism and dN/dS. However, it is difficult to assess this explanation without an independent way of estimating N_e for these species.

The two explanations just discussed both concern the role of selection in shaping mutation rates, whether increased positive selection for novelty or less efficient negative selection on DNA copy fidelity and repair. In addition, there are many other species traits that can influence mutation rates, so if these traits differ consistently between parasites and their nonparasitic relatives, it could potentially explain the pattern in rates of molecular evolution. The most obvious candidate is number of DNA replications per unit time.

Species with shorter generation times tend to have higher rates of molecular evolution, presumably because they accumulate more

DNA replication errors per unit time [51]. In plants, it has been widely reported that annual plants and herbs have faster substitution rates than perennials and woody plants, respectively [52], though it is not clear exactly how this relates to number of genome copies per unit time [29]. Plant height also correlates with rate of molecular evolution in plants, which may be because taller plants tend to have longer generation times (and so fewer meiosis per unit of time) and slower average growth rates (and so fewer mitoses per unit of time) [53]. Highly reduced holoparasites that do not produce stems, such as *Pilostyles* and *Apodanthes,* are clearly much shorter than their freeliving relatives, such as *Hibiscus* and*Theobroma* Similarly, the fungus-like *Cynomorium coccineum* is much smaller than its nonparasitic relative *Peridiscus lucidus*, a South American tree species. So if parasitic plants tend to have smaller bodies, shorter lives or more rapid generation turnover than their nonparasitic relatives, then they might also have higher mutation rates because their genomes will undergo more replications per unit time.

Although we do not have reliable measures of average height of all species in our comparisons, we can make a broad assessment based on growth form. In the majority of comparisons, the parasitic species are clearly smaller than the nonparasitic species – specifically in the Apodanthaceae (comparison 1), Cytinaceae (2), Rafflesiaceae (3), Cynomoriaceae (4), Mitrastemonaceae (6), Lauraceae (10), Hydnoraceae (11), and the clade consisting of the parasitic relatives of the Olacaceae (12). These comparisons all show higher rates in the parasitic species for the nuclear genes and mitochondrial genes except for comparison 6 (Mitrastemonaceae) which has slower rates for the parasitic species for mitochondrial dS and chloroplast 16S rRNA.

While it provides a plausible explanation of the patterns we observe, the plant height hypothesis does not provide a perfect fit to all of the substitution rate differences in this study. For example, the perennial hemiparasite *Krameria lanceolata* has lower nuclear rates and mitochondrial dS than the annual herb *Kallstroemia parviflora,* which is consistent with larger size in the parasitic taxon, but it also has faster mitochondrial dN. For the chloroplast comparison, *Krameria* has lower substitution rates than its larger relatives, the tree *Guaiacum sanctum* and the woody shrub *Larrea cuneifolia*. For the remaining two comparisons (9 and 10), there is not a clear difference in body size, but the parasites have the higher substitution rates in all alignments. So while the patterns we observe are broadly consistent with an effect of plant size on rates of mutation rates, the match to the observations is not exact, and there may be other factors at play.

Another potential source of an indirect link between molecular evolution and parasitic lifestyle is through environmental effects. It has been suggested that plant lineages at low latitudes have increased rates of molecular evolution, potentially due to the mutagenic effect of UV radiation, or to higher growth rates and faster generation turnover [53-55]. But latitude or environmental energy is unlikely to provide an explanation for higher substitution rates in parasitic plants. Some parasitic plant groups are primarily tropical, such as the Rafflesiaceae, found in South East Asia. But most of the parasitic lineages included in this study have a broader distribution, including both temperate and tropical species, for example the Orobanchaceae (broom rapes) and *Cuscuta*(dodder). Similarly, net diversification rate in plants has been linked to both synonymous and nonsynonymous rates [56, 57], but since parasitic plant clades tend to be less diverse than their nonparasitic sister clades [58], this seems unlikely to provide an explanation for our results.

CONCLUSIONS

The study of parasitic plants has many important outcomes. By reducing the water and nutrients available to host plant growth, parasitic plants have a serious impact on agriculture, which is estimated to impact on the food supply of over 100 million people [27]. Parasitic plants also provide an excellent case study for examining many evolutionary processes, including diversification and specialization, genome evolution, and host-parasite interactions. Demonstrating consistently higher substitution rates in parasitic plants not only provides a window on molecular evolution, it also has practical implications for the use of DNA sequence data in evolutionary and ecological research. Molecular markers have played a key role not only in establishing the phylogeny and systematics of parasitic plants, but in understanding broader evolutionary patterns, such specialization to different host plants. Consistent differences in rate of molecular evolution in parasitic lineages should be considered when interpreting molecular phylogenies and date estimates.

METHODS

Data

We gathered DNA sequence data from GenBank for each of the parasitic plant clades identified by Barkman *et al.* [17]. We focus on angiosperms, so we do not include the gymnosperm *Parasitaxus,*nor the liverwort *Aneura mirabilis,* both of which appear to form parasitic attachments to host plants via a fungal partner [20,59]. Nor do we include mycoheterotrophic angiosperm species, such as those in the Ericaceae [60], which derive part of their nutrition through a relationship with fungi, because the line between fully parasitic mycoheterotrophs and plants that have mycorrhizal partners is not always clear [61,62].

We use a sister-clades approach, identifying clades of parasitic plants and their autotrophic (non-parasitic) relatives. Each lineage of the sister clade has had the same amount of time since their last common ancestor to accumulate substitutions, therefore any difference in branch length (estimated number of substitutions since the last common ancestor) is likely to reflect a difference in the net substitution rate [16,56]. If parasitic plants have a higher rate of molecular evolution, then we would expect to see a longer total branch length in the parasitic clade compared to its non-parasitic sister clade [51].

The position of parasitic clades within angiosperm phylogeny has been controversial, and even the higher-level relationships are debated [63-65]. As a consequence, there have been many changes to higher-level taxonomy. For simplicity, we report the GenBank taxonomy specified for each sequence used.

We identified close non-parasitic relatives of each of the parasitic lineages, using published phylogenies and taxonomies as a guide [17, 64-72]. For the purposes of this study, we do not need to know the exact phylogenetic position of parasitic taxa, as long as we can be sure that we are comparing phylogenetically independent origins of parasitic lineages to their autotrophic relatives. For each independently evolved parasitic lineage we aimed to select a related autotrophic clade that could be placed outside the parasitic clade with certainty, so we did not always choose the closest relative where phylogeny was uncertain. This gave us 12 phylogenetically independent comparisons between parasitic and non-parasitic lineages (Tables 1and 2). Each parasitic and non-parasitic sister clade was represented by sequences from up to nine different species we always balanced the number of species in the sister clades, to minimize the effects of node density on substitution rate estimation [16].

Table 1: Estimates of branch length for sister clades of parasitic (P) and autotrophic (non-parasitic, NP) plants for nuclear and mitochondrial sequences

Estimates of branch length for sister clades of parasitic (P) and autotrophic (non-parasitic, NP) plants for nuclear and mitochondrial sequences

Comparison	Parasitic (P)	Non-parasite (NP)	Nuclear			Mitchondrial (all substitutions)			Mitchondrial (dN)			Mitchondrial (dS)			Mitchondrial (dN/dS)		
			P	NP	Sign	P	NP	Sign	P	NP	Sign	P	NP	Sign	P	NP	Sign
1	Apodanthaceae	Malvaceae	0.513	0.072	+	0.380	0.037	+	0.466	0.055	+	0.739	0.024	+	0.628	1.287	-
2	Cytinaceae	Thymelaeaceae	0.121	0.028	+	0.670	0.035	+	0.077	0.032	+	0.094	0.043	+	0.960	0.742	+
3	Rafflesiaceae	Passifloraceae/ Euphorbiaceae	1.093	0.052	+	0.631	0.141	+	0.577	0.114	+	0.660	0.241	+	0.726	0.520	+
4	Cynomoriaceae	Hamamelidaceae/ Peridiscaceae	0.087	0.012	+	0.030	0.011	+	0.017	0.007	+	0.060	0.024	+	0.276	0.282	-
5	Krameriaceae	Zygophyllaceae	0.019	0.025	-	0.015	0.011	+	0.012	0.007	+	0.023	0.025	-	0.529	0.289	+
6	Mitrastemonaceae	Vacciniae	0.075	0.013	+	0.026	0.017	+	0.021	0.005	+	0.039	0.048	-	0.529	0.108	+
7	Boraginaceae	Boraginaceae	0.036	0.028	+	0.029	0.020	+	0.020	0.013	+	0.049	0.044	+	0.389	0.343	+
8	Orobanchaceae/ Rhinantheae	Lamiales/ Plantaginaceae	0.040	0.043	-	0.007	0.022	-	0.002	0.006	-	0.023	0.072	-	0.133	0.054	+
9	Convolvulaceae	Ipomoeeae	0.080	0.016	+	0.061	0.021	+	0.052	0.008	+	0.096	0.054	+	0.562	0.249	+
10	Lauraceae	Lauraceae	0.323	0.230	+	0.012	0.005	+	0.010	0.004	+	0.018	0.008	+	0.540	0.530	+
11	Hydnoraceae	Aristolochiaceae	0.296	0.057	+	0.152	0.042	+	0.072	0.014	+	0.152	0.042	+	0.418	0.330	+
12	Balanophoraceae/Loranthceae/ Schoepfiaceae/ Olacaceae	Olacaceae	0.384	0.120	+	0.131	0.014	+	0.101	0.005	+	0.180	0.040	+	0.493	0.773	-
Wilcoxon signed-ranks test			Z = 2.786, p = 0.005			Z = 2.628, p = 0.009			Z = 2.942, p = 0.003			Z = 2.157, p = 0.031			Z = 1.294, p = 0.196		

For details of the species and sequences included in each comparisons see Mitochondrial branch lengths were estimated for the whole alignment (all substitutions), as well as no synonymous (dN) and synonymous substitutions (dS): see methods for details. Sign indicates whether the parasite has the longer branch length (+) or not (−). The Wilcoxon ranked signs test takes into account the magnitude of differences in branch length as well as the sign.

Bromham *et al.* *BMC Evolutionary Biology* 2013 13:126 doi: 10.1186/1471-2148-13-126

Table 2: Estimates of branch length for sister clades of parasitic (P) and autotrophic (non-parasitic, NP) plants for chloroplast sequences

Estimates of branch length for sister clades of parasitic (P) and autotrophic (non-parasitic, NP) plants for chloroplast sequences

Comparison	Parasitic clade	Non-parasite	16S rRNA (*rrn16*)			Protein-coding (all)			Comb.	dN			dS			dN/dS		
			P	NP	Sign	P	NP	Sign	Sign	P	NP	Sign	P	NP	Sign	P	NP	Sign
1	Apodanthaceae	Malvaceae	1.637	0.013	+				+									
2	Cytinaceae	Thymelaeaceae	0.091	0.008	+				+									
3	Rafflesiaceae	Passifloraceae/ Euphorbiaceae	0.150	0.091	+				+									
4	Cynomoriaceae	Hamamelidaceae/ Penthoraceae	0.083	0.002	+				+									
5	Krameriaceae	Zygophyllaceae				0.098	0.164	-	-	0.032	0.070	-	0.296	0.360	-	0.112	0.220	-
6	Mitrastemonaceae	Vaccinieae	0.196	0.940	-				-									
7	Boraginaceae	Boraginaceae				0.076	0.019	+	+	0.040	0.017	+	0.133	0.040	+	0.297	0.423	-
8	Orobanchaceae	Plantaginaceae	0.027	0.001	+	0.561	0.366	+	+	0.413	0.206	+	0.960	0.743	+	0.409	0.268	+
9	Convolvulaceae	Convolvulaceae/ Ipomoeeae	0.014	0.001	+	0.724	0.052	+	+	0.115	0.027	+	1.938	0.151	+	0.047	0.102	-
10	Lauraceae	Lauraceae				0.344	0.122	+	+	0.235	0.116	+	0.526	0.137	+	0.320	0.721	-
11	Hydnoraceae	Aristolochiaceae	0.336	0.000	+				+									
12	Santalaceae/ Olacaceae	Strombosiaceae				0.125	0.049	+	+	0.068	0.039	+	0.241	0.069	+	0.239	0.585	-
Wilcoxon rank sign test			Z = 1.47, p = 0.14			Z = 1.68, p = 0.09				Z = 1.47, p = 0.14			Z = 1.89, p = 0.06			Z = −1.26, p = 0.21		
Signs test									p = 0.039									

For details of the species and sequences included in each comparisons see Protein-coding branch lengths were estimated for all substitutions (all), nonsynonymous (dN) and synonymous substitutions (dS): see methods for details. Sign indicates whether the parasite has the longer branch length (+) or not (−). The Wilcoxon ranked signs test takes into account the magnitude of differences in branch length as well as the sign. Because most comparisons had only 16S or the protein-coding genes, we combined the two (Comb.) to assess the pattern over all comparisons using a signs test: for this test we considered only the direction of the rate difference for each comparison, whether calculated from 16S or protein coding genes or both.

Bromham *et al. BMC Evolutionary Biology* 2013 13:126 doi: 10.1186/1471-2148-13-126

We targeted genes for which sequences were available for the largest number of the parasitic taxa and their non-parasitic relatives: maturase R (*matR*), cytochrome oxidase subunit I (*coxI*), NADH dehydrogenase subunit 1 (*nad1*) and ATPase F1 alpha subunit (*atp1*) from the mitochondrial genome; 18S small subunit ribosomal RNA (*rrn18*) and 26S ribosomal RNA (*rrn26*) from the nuclear genome; ribulose-1, 5-bisphosphate carboxylase/oxygenase large subunit (*rbcL*), maturase K (*matK*) and 16S ribosomal RNA (*rrn16*) from the chloroplast genome. We used the GeneFinder script [73] to find the longest available sequences for each gene on GenBank. Some gene sequences were unavailable for some taxa (in particular, chloroplast sequences were often not available for holoparasites). A complete list of the sequences, along with their GenBank accession numbers, is available in the Supplementary Material

Analysis

We aligned sequences in Geneious version 6.0 [74]. We aligned protein-coding exons in frame, using the amino acid translations as a guide, with the MUSCLE translation alignment plugin [75], and we aligned non-coding sequences using the standard MUSCLE alignment plugin in Geneious. We then adjusted all alignments by eye. When we identified pseudogenes in chloroplast protein sequences we removed them from the alignments. We produced three separate genome alignments by concatenating the genes from each genome (mitochondrial, chloroplast, and nuclear). For each genome alignment, substitution models and partitioning schemes were chosen using AICc in Partition Finder [76]. The best substitution models and partition schemes were used for phylogenetic estimation in RAxML v7.0.4 [77].

Separate maximum likelihood analyses were run for each alignment.

Next we created a single maximum likelihood (ML) tree topology and corresponding alignment for each of the three genomes (nuclear, plastid, and mitochondrial) for each of the 12 sister pairs (36 topologies and alignments in total). For each comparison, we then extracted the parasitic and non-parasitic taxa plus a closely related non-parasite outgroup from a different order from each of the genome trees using the APE package [78], and produced corresponding alignment files. For each alignment, we only included gene regions that had coverage for both the parasitic and non-parasitic clades. We removed the small introns of the *coxI* and *nad1* genes in the mitochondrial alignment.

We estimated total branch length for all alignments using maximum likelihood and a GTR + G model in the baseml program of the PAML package [79]. For protein-coding genes (mitochondrial *matR,coxI, nad1* and *atp1;* chloroplast *rbcL* and *matK*), we estimated synonymous, and nonsynonymous branch lengths using the GY94 codon substitution model [80] in the codeml program of the PAML package [79], with dN/dS values free to vary across the tree. In addition, we estimated clade-specific dN/dS ratios ( ) for the parasitic and non-parasitic clade in each comparison.

To compare the rate of accumulation of substitutions in parasitic lineages to their nonparasitic sister lineage, we calculated the total branch length of the parasitic and non-parasitic clades in each sister pair using a python script. Total branch lengths for ribosomal RNA genes and protein-coding genes, and total dS and dN branch lengths for protein-coding genes were calculated by summing all edge lengths for each clade from the shared node with its sister clade. Since estimates of substitution rates can be influenced by base composition bias, we also calculated GC content for each sister clade. We tested for a significant association between parasitism and rate of molecular evolution using a Wilcoxon signed ranks test in the SPlus package v8.2.

AUTHORS' CONTRIBUTIONS

All authors contributed to data collection and analysis and manuscript preparation and all authors read and approved the final manuscript.

ACKNOWLEDGEMENTS

We thank Katherine Byron, Simon Ho and Art Wuster for their contributions to earlier versions of this project.

REFERENCES

1. Haraguchi Y, Sasaki A: Host-parasite arms-race in mutation modifications - indefinate escalation despite a heavy load? *J Theor Biol* 1996, 183(2):121-137.
2. Pal C, Maciá MD, Oliver A, Schachar I, Buckling A: Coevolution with viruses drives the evolution of bacterial mutation rates. *Nature* 2007, 450(7172):1079-1081.
3. Paterson S, Vogwill T, Buckling a, Benmayor R, Spiers AJ, Thomson NR, Quail M, Smith F, Walker D, Libberton B: Antagonistic coevolution accelerates molecular evolution. *Nature* 2010, 464(7286):275-278.
4. Young ND, De Pamphilis CW: Rate variation in parasitic plants: correlated and uncorrelated patterns among plastid genes of different function. *BMC Evol Biol* 2005, 5:16
5. Dowton M, Austin AD: Increased genetic diversity in mitochondrial genes is correlated with the evolution of parasitism in the Hymenoptera. *J Mol Evol* 1995, 41:958-965
6. Duff RJ, Nickrent DL: Characterization of mitochondrial small-subunit riibosomal RNAs from holoparasitic plants. *J Mol Evol* 1997, 45(6):631
7. Lemaire B, Huysmans S, Smets E, Merckx V: Rate accelerations in nuclear 18S rDNA of mycoheterotrophic and parasitic angiosperms. *J Plant Res* 2011, 124(5):561-576.
8. Jiggins FM, Hurst GDD, and Yang Z: Host-symbiont conflicts: positive selection on an outer membrane protein of parasitic but not mutualistic Rickettsiaceae. *Mol Biol Evol* 2002, 19:1341-1349
9. Kuijt J: *The biology of flowering plants, vol.* Berkeley: University of California Press; 1969.

10. Westwood JH, Yoder JI, Timko MP, De Pamphilis CW: The evolution of parasitism in plants. *Trends Plant Sci* 2010, 15(4):227-235
11. Nickrent DL, Starr EM: High rates of nucleotide substitution in nuclear small-subunit (18S) rDNA from holoparasitic flowering plants. *J Mol Evol* 1994, 39(1):62.
12. Nickrent DL, Garcia MA: On the brink of holoparasitism: plastome evolution in dwarf mistletoes (Arceuthobium, Viscaceae). *J Mol Evol* 2009, 68(6):603-615.
13. Su H-J, Hu J-M: Rate heterogeneity in six protein-coding genes from the holoparasite*Balanophora* (Balanophoraceae) and other taxa of Santalales. *Ann Bot* 2012, 110:1137-1147.
14. McNeal JR, Kuehl JV, Boore JL, De Pamphilis CW: Complete plastid genome sequences suggest strong selection for retention of photosynthetic genes in the parasitic plant genus *Cuscuta*. *BMC Plant Biol* 2007, 7(1):57.
15. Wolfe KH, Katz-Downie DS, Morden CW, Palmer JD: Evolution of the plastid ribosomal RNA operon in a nongreen parasitic plant: accelerated sequence evolution, altered promoter structure, and tRNA pseudogenes. *Plant Mol Biol* 1992, 18(6):1037-1048
16. Lanfear R, Welch JJ, Bromham L: Watching the clock: Studying variation in rates of molecular evolution. *Trends Ecol Evol* 2010, 25(9):495-503.
17. Barkman TJ, McNeal JR, Lim SH, Coat G, Croom HB, Young ND, De Pamphilis CW:Mitochondrial DNA suggests at least 11 origins of parasitism in angiosperms and reveals genomic chimerism in parasitic plants. *BMC Evol Biol* 2007, 7:248.
18. Nickrent DL: *Parasitic plants connection*. Southern Illinois: Southern Illinois University; 1997.
19. Krause K: From chloroplasts to cryptic plastids: evolution of plastid genomes in parasitic plants. *Curr Genet* 2008, 54(3):111-121.
20. Wickett NJ, Zhang Y, Hansen SK, Roper JM, Kuehl JV, Plock SA, Wolf PG, DePamphilis CW, Boore JL, Goffinet B: Functional gene losses occur with minimal size reduction in the plastid genome of the parasitic liverwort *Aneura mirabilis*. *Mol Biol Evol* 2008, 25(2):393-401

21. Randle CP, Wolfe AD: The evolution and expression of rbcL in holoparasitic sister-genera*Harveya* and *Hyobanche* (Orobanchaceae). *Am J Bot* 2005, 92(9):1575-1585
22. Young ND, DePamphilis CW: Purifying selection detected in the plastid gene matK and flanking ribozyme regions within a group II intron of nonphotosynthetic plants. *Mol Biol Evol* 2000, 17(12):1933-1941
23. Thorogood CJ, Rumsey FJ, Harris SA, Hiscock SJ: Host-driven divergence in the parasitic plant *Orobanche minor* Sm. (Orobanchaceae). *Mol Ecol* 2008, 17(19):4289-4303
24. Van der Kooij T, Krause K, Dörr I, Krupinska K: Molecular, functional and ultrastructural characterisation of plastids from six species of the parasitic flowering plant genus*Cuscuta*. *Planta* 2000, 210(5):701-707
25. Funk H, Berg S, Krupinska K, Maier U, Krause K: Complete DNA sequences of the plastid genomes of two parasitic flowering plant species, Cuscuta reflexa and Cuscuta gronovii. *BMC Plant Biol* 2007, 7(1):45
26. Barbrook AC, Howe CJ, Purton S: Why are plastid genomes retained in non-photosynthetic organisms? *Trends Plant Sci* 2006, 11(2):101-108
27. Yoder JI, Scholes JD: Host plant resistance to parasitic weeds; recent progress and bottlenecks. *Curr Opp Plant Biol* 2010, 13(4):478-484
28. Honaas L, Wafula E, Yang Z, Der J, Wickett N, Altman N, Taylor C, Yoder J, Timko M, and Westwood J, *et al.*: Functional genomics of a generalist parasitic plant: Laser microdissection of host-parasite interface reveals host-specific patterns of parasite gene expression. *BMC Plant Biol* 2013, 13(1):9
29. Gaut B, Yang L, Takuno S, Eguiarte LE: The patterns and causes of variation in plant nucleotide substitution rates. *Annu Rev Ecol Evol Syst* 2011, 42:245-266.
30. Gu WJ, Wang XF, Zhai CY, Xie XY, Zhou T: Selection on synonymous sites for increased accessibility around miRNA binding sites in plants. *Mol Biol Evol* 2012, 29(10):3037-3044
31. Qiu S, Bergero R, Zeng K, Charlesworth D: Patterns of codon usage bias in *Silene latifolia*. *Mol Biol Evol* 2011, 28(1):771-780

32. Lynch M: *The origins of genome architecture*. Sunderland, Mass: Sinauer Assoc; 2007.
33. Woolfit M: Effective population size and the rate and pattern of nucleotide substitutions. *Biol Lett* 2009, 5(3):417-420
34. Charlesworth B: Effective population size and patterns of molecular evolution and variation. *Nature Rev Genet* 2009, 10:195-205
35. Ebert D: Experimental evolution of parasites. *Science* 1998, 282(5393):1432-1436.
36. Woolfit M, Bromham L: Increased rates of sequence evolution in endosymbiotic bacteria and fungi with small effective population sizes. *Mol Biol Evol* 2003, 20(9):1545-1555
37. Musselman LJ: The biology of *Striga, Orobanche,* and other root-parasitic weeds. *Annu Rev Phytopathol* 1980, 18(1):463-489
38. Lahti DC, Johnson NA, Ajie BC, Otto SP, Hendry AP, Blumstein DT, Coss RG, Donohue K, Foster SA: Relaxed selection in the wild. *Trends Ecol Evol* 2009, 24(9):487-496
39. Hafner M, Sudman P, Villablanca F, Spradling T, Demastes J, Nadler S: Disparate rates if molecular evolution in cospeciating hosts and parasites. *Science* 1994, 265:1087-1090
40. Kaltz O, Shykoff JA: Local adaptation in host,Àìparasite systems. *Heredity* 1998, 81(4):361-370
41. Lanfear R, Ho SYW, and Love D, Bromham L: Mutation rate influences diversification rate in birds. *Proc natl Acad Sci USA* 2010, 107(47):20423-20428
42. Lartillot N: Interaction between Selection and Biased Gene Conversion in Mammalian Protein-Coding Sequence Evolution Revealed by a Phylogenetic Covariance Analysis. *Mol Biol Evol* 2013, 30(2):356-368
43. DeRose-Wilson L, Gaut B: Transcription-related mutations and GC content drive variation in nucleotide substitution rates across the genomes of Arabidopsis thaliana and Arabidopsis lyrata. *BMC Evol Biol* 2007, 7(1):66
44. Taddei F, Radman M, Maynard-Smith J, Toupance B, Gouyon P, Godelle B: Role of mutator alleles in adaptive evolution. *Nature* 1997, 387(6634):700-702

45. Giraud A, Matic I, Tenaillon O, Clara A, Radman M, Fons M, Taddei F: Costs and benefits of high mutation rates: adaptive evolution of bacteria in the mouse gut. *Science* 2001, 291(5513):2606-2608
46. Metzgar D, Wills C: Evidence for the adaptive evolution of mutation rates. *Cell* 2000, 101(6):581
47. Rainey PB: Evolutionary genetics: the economics of mutation. *Curr Biol* 1999, 9(10):R371-R373
48. Denamur E, Tenaillon O, Deschamps C, Skurnik D, Ronco E, Gaillard JL, Picard B, Branger C, Matic I: Intermediate mutation frequencies favor evolution of multidrug resistance in*Escherichia coli*. *Genetics* 2005, 171(2):825-827
49. Weissman DB, Barton NH: Limits to the rate of adaptive substitution in sexual populations. *PLoS Genet* 2012, 8(6):e1002740
50. Lynch M: The Lower Bound to the Evolution of Mutation Rates. *Genome Biol Evol* 2011, 3:1107-1118
51. Bromham L: Why do species vary in their rate of molecular evolution? *Biol Lett* 2009, 5:401-404
52. Smith SA, Donoghue MJ: Rates of molecular evolution are linked to life history in flowering plants. *Science* 2008, 322:86-89
53. Lanfear R, Ho SYW, Davies TJ, Moles ATA L, Swenson NG, Warman L, Zanne AE, Allen AP:Taller plants have lower rates of molecular evolution: the rate of mitosis hypothesis. *Nature Communications* 2013, 4:1879
54. Davies TJ, Savolainen V, Chase MW, Moat J, Barraclough TG: Environmental energy and evolutionary rates in flowering plants. *Proc R Soc Lond B* 2004, 271:2195-2220.
55. Wright SD, Keeling J, Gillman L: The road from Santa Rosalia: a faster tempo of evolution in tropical climates. *Proc natl Acad Sci USA* 2006, 103:7718-7722.
56. Barraclough TG, Savolainen V: Evolutionary rates and species diversity in flowering plants. *Evolution* 2001, 55:677-683
57. Duchene D, Bromham L: Rates of molecular evolution and diversification in plants: chloroplast substitution rates correlate with species richness in the Proteaceae. *BMC Evol Biol* 2013, 13:65

58. Hardy NB, Cook LG: Testing for Ecological Limitation of Diversification: A Case Study Using Parasitic Plants. *Am Nat* 2012, 180(4):438-449
59. Feild TS, Brodribb TJ: A unique mode of parasitism in the conifer coral tree Parasitaxus ustus (Podocarpaceae). *Plant Cell Environ* 2005, 28(10):1316-1325.
60. Braukmann T, Stefanovicá S: Plastid genome evolution in mycoheterotrophic Ericaceae. *Plant Mol Biol* 2012, 1:16.
61. Leake JR, Cameron DD: Physiological ecology of mycoheterotrophy. *New Phytol* 2010, 185(3):601-605
62. Merckx V, Freudenstein JV: Evolution of mycoheterotrophy in plants: a phylogenetic perspective. *New Phytol* 2010, 185(3):605-609
63. Filipowicz N, Renner SS: The worldwide holoparasitic Apodanthaceae confidently placed in the Cucurbitales by nuclear and mitochondrial gene trees. *BMC Evol Biol* 2010, 10(1):219
64. Nickrent DL, Der JP, Anderson FE: Discovery of the photosynthetic relatives of the "Maltese mushroom" *Cynomorium*. *BMC Evol Biol* 2005, 5(1):38
65. Nickrent D, Blarer A, Qiu Y-L, Vidal-Russell R, and Anderson F: Phylogenetic inference in Rafflesiales: the influence of rate heterogeneity and horizontal gene transfer. *BMC Evol Biol* 2004, 4:1471-2148.
66. Barkman TJ, Lim S-H, Salleh KM, Nais J: Mitochondrial DNA sequences reveal the photosynthetic relatives of *Rafflesia,* the world's largest flower. *Proc natl Acad Sci USA* 2004, 101:787-792.
67. Vr M, Nickrent DL: Molecular phylogenetic relationships of Olacaceae and related Santalales. *Syst Bot* 2008, 33(1):97-106
68. Olmstead RG, Wolfe AD, Young ND, Elisons WJ, Reeves PA: Disintegration of the Scrophulariaceae. *Am J Bot* 2001, 88(2):348-361
69. Rohwer JG: Toward a phylogenetic classification of the Lauraceae: evidence from matK sequences. *Syst Bot* 2000, 25(1):60-71
70. Sheahan M, Chase M: A phylogenetic analysis of Zygophyllaceae R. Br. based on morphological, anatomical and rbcL DNA sequence data. *Bot J Linn Soc* 1996, 122:279-300.

71. Soltis D, Soltis P, Chase M, Mort M, Albach D, Zanis M, Savolainen V, Hahn W, and Hoot S, Fay M: Angiosperm phylogeny inferred from 18s rDNA, rbcL, and atpB sequences. *Bot J Linn Soc* 2000, 133:381-461.

72. Stefanovic S, Olmstead RG: Testing the phylogenetic position of a parasitic plant (*Cuscuta*, Convolvulaceae, Asteridae): Bayesian inference and the parametric bootstrap on data drawn from three genomes. *Syst Biol* 2004, 53(3):384-399.

73. Lanfear R, Bromham L: Estimating phylogenies for species assemblages: a complete phylogeny for the past and present native birds of New Zealand. *Mol Phylog Evol* 2011, 61:958-963

74. Drummond AJ, Ashton B, Buxton S, Cheung M, Cooper A, Duran C, Field M, Heled J, Kearse M, Markowitz S, Moir R, *et al.*: *Geneious*. 2011

75. Edgar RC: MUSCLE: multiple sequence alignment with high accuracy and high throughput. *Nucleic Acids Res* 2004, 32:1792-1797

76. Lanfear R, Calcott B, and Ho SYW, Guindon S: PartitionFinder: combined selection of partitioning schemes and substitution models for phylogenetic analyses. *Molecular Biology and Evolution* 2012, 29(6):1695-1701.

77. Stamatakis A: RAxML-VI-HPC: maximum likelihood-based phylogenetic analyses with thousands of taxa and mixed models. *Bioinformatics* 2006, 22(21):2688-2690.

78. Paradis E, Claude J: K. S: APE: analyses of phylogenetics and evolution in R language. *Bioinformatics* 2004, 20:289-290.

79. Yang Z: PAML: a program package for phylogenetic analysis by maximum likelihood. *Mol Biol Evol* 2007, 24:1586-1591. http://abacus.gene.ucl.ac.uk/software/paml.html

80. Goldman N, Yang Z: A codon-based model of nucleotide substitution for protein-coding DNA sequences. *Mol Biol Evol* 1994, 11:725-736.

Chapter 3

A Molecular Timescale of Eukaryote Evolution and the Rise of Complex Multicellular Life

S Blair Hedges[1], Jaime E Blair[1], Maria L Venturi[1], and Jason L Shoe[2]

[1]NASA Astrobiology Institute and Department of Biology, the Pennsylvania State University University Park, PA 16802, USA

[2]Harvard School of Dental Medicine, 188 Longwood Avenue, Boston, MA 02115, USA

ABSTRACT

Background

The pattern and timing of the rise in complex multicellular life during Earth's history has not been established. Great disparity persists between the pattern suggested by the fossil record and that estimated by molecular clocks, especially for plants, animals, fungi, and the deepest branches of the eukaryote tree. Here, we used all available protein sequence data and molecular clock methods to place constraints on the increase in complexity through time.

Results

Our phylogenetic analyses revealed that (i) animals are more closely related to fungi than to plants, (ii) red algae are closer to plants than to animals or fungi, (iii) choanoflagellates are closer to animals than to fungi or plants, (iv) diplomonads, euglenozoans, and alveolates each are basal to plants+animals+fungi, and (v) diplomonads are basal to other eukaryotes (including alveolates and euglenozoans). Divergence times were estimated from global and local clock methods using 20–188 proteins per node, with data treated separately (multigene) and concatenated (supergene). Different time estimation methods yielded similar results (within 5%): vertebrate-arthropod (964 million years ago, Ma), Cnidaria-Bilateria (1,298 Ma), Porifera-Eumetozoa (1,351 Ma), Pyrenomycetes-Plectomycetes (551 Ma), *Candida-Saccharomyces* (723 Ma), Hemiascomycetes-filamentous Ascomycota (982 Ma), Basidiomycota-Ascomycota (968 Ma), Mucorales-Basidiomycota (947 Ma), Fungi-Animalia (1,513 Ma), mosses-vascular plants (707 Ma), Chlorophyta-Tracheophyta (968 Ma), Rhodophyta-Chlorophyta+Embryophyta (1,428 Ma), Plantae-Animalia (1,609 Ma), Alveolata-plants+animals+fungi (1,973 Ma), Euglenozoa-plants+animals+fungi (1,961 Ma), and Giardia-plants+animals+fungi (2,309 Ma). By extrapolation, mitochondria arose approximately 2300-1800 Ma and plastids arose 1600-1500 Ma. Estimates of the maximum number of cell types of common ancestors, combined with divergence times, showed an increase from two cell types at 2500 Ma to ~10 types at 1500 Ma and 50 cell types at ~1000 Ma.

Conclusions

The results suggest that oxygen levels in the environment, and the ability of eukaryotes to extract energy from oxygen, as well as produce oxygen, were key factors in the rise of complex multicellular life. Mitochondria and organisms with more than 2–3 cell types appeared soon after the initial increase in oxygen levels at 2300 Ma. The addition of plastids at

1500 Ma, allowing eukaryotes to produce oxygen, preceded the major rise in complexity.

BACKGROUND

Organismal complexity can be defined in many ways, although the most common measure is the number of cell types [1-4]. Prokaryotes and many unicellular eukaryotes have only one or a few cell types, but vertebrates have more than 100 [1]. If cell types provide a tracer of complex life, it is of interest to know the general pattern of increase over the history of life. For example, a literal interpretation of the Cambrian explosion (520 million years ago, Ma), when many animal phyla first appeared in the fossil record, would be that a rapid increase in complexity occurred during the last one-ninth of the history of the planet. This apparent delay in the evolution of complex life on Earth has contributed to the argument that complex life may be rare in the universe [5]. Molecular clocks have yielded earlier times for the origin of animal phyla [6-9], but the methods have received criticism [10,11]. At the same time, recent fossil discoveries have pushed back the origins of some groups of eukaryotes [12,13], although a great discordance remains between most molecular clock results and the fossil record.

In this study, we have estimated a contour for the rise in complex life using a phylogeny and timescale derived from currently available protein sequence data. Ancestral numbers of cell types were estimated using the resulting phylogenetic and temporal framework. We have taken care to address criticisms of past molecular clock studies and have used all available timing methods applicable to protein sequence data, including global (constant rate) and local (variable rate) methods. The methods include those based on least-squares analysis [14], Bayesian inference[15], and penalized likelihood [16]. To avoid any potential artifacts arising from analysis of multiple alignments [17,18], we have also used concatenated datasets [19]. We have tested our calibrations for reciprocity [20] and have used both vertebrate and non-vertebrate fossil calibrations and constraints. The results support a deep history for complex multicellular eukaryotes, and implicate oxygen as a possible trigger for the rise in complex life.

RESULTS

Phylogenetic Analyses

Our analyses of the concatenated data sets produced the following results: (i) animals are more closely related to fungi than to plants, (ii) red algae are closer to plants than to animals or fungi, (iii) choanoflagellates are closer to animals than to fungi or plants, (iv) diplomonads, euglenozoans, and alveolates each are basal to plants+animals+fungi, and (v) diplomonads are basal to other eukaryotes (including alveolates and euglenozoans) (Fig. 1). Most of these relationships are uncontroversial except for the uncertain position of the root of the tree as discussed elsewhere [21]. Our results with nuclear proteins agree with earlier ribosomal RNA trees [22] in supporting a root near the excavates (e.g., diplomonads) rather than on the opisthokont-amoebozoan branch (e.g., animals, fungi, and amoebas) [23]. Confidence values for these relationships were high (>99%) using three phylogenetic methods (maximum likelihood, minimum evolution, and Bayesian inference) in five of the seven analyses (Fig. 1). For the remaining two analyses (ii and v), significant support values were obtained with Bayesian inference, but varied for maximum likelihood and minimum evolution.

Protenis:	151	19	6
Amino acids:	75,287	5,391	3,195
Maximum likelihood:	100%	64%	100%
Minimum evolution:	100%	85%	100%
Bayesian inference:	100%	100%	100%

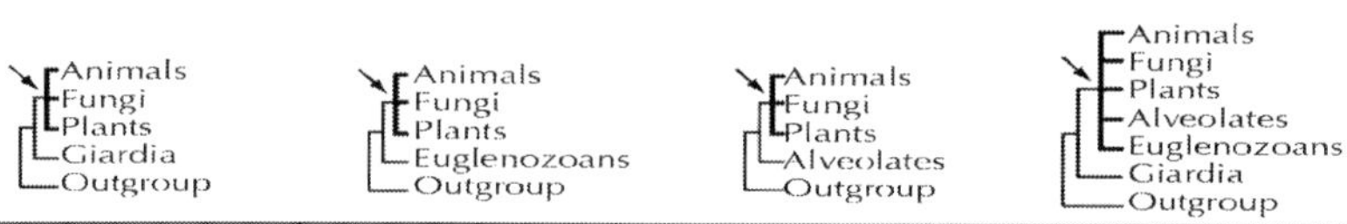

39	105	73	4%
14,473	38,492	27,497	2,103
100%	100%	100%	92%
100%	100%	100%	98%
100%	100%	100%	100%

Figure 1: Phylogenetic relationships of selected eukaryotes. For each data set (column), all taxa are represented in all proteins. Support values are listed

for the three methods (maximum likelihood, minimum evolution, Bayesian inference) and correspond to the node indicated by the arrow (and bolded group) for each tree.

Divergence Times Estimated with Different Methods

We estimated three deep (Precambrian) divergences in the eukaryote tree using the primary (bird-mammal) calibration and MGL_{LS} (see Methods). In each case, there were no missing data; the data sets contained all proteins for all taxa. The divergence times were: vertebrate-arthropod (964 ± 132 Ma; 151 total and 120 rate constant proteins; 49,644 amino acids), animal-fungi (1492 ± 46 Ma; 188 total and 89 rate constant proteins; 31,362 amino acids), and animal-plant (1524 ± 53 Ma; 188 total and 143 rate constant proteins; 60,274 amino acids) (Table 1). These dates were similar to previous estimates using fewer proteins and different methods [8], and as secondary calibration points were found to be consistent in tests of reciprocity (see next section). In turn, these three time estimates were used as calibrations for estimating other divergence times using least-squares and penalized likelihood methods, and the 95% confidence intervals were used as nodal constraints for the Bayesian analysis. Rate parameters and a list of proteins used in the analyses are in supplemental tables 1, 2.

Table 1: Estimates of divergence time among eukaryotes

Node[a]	Comparison	Proteins[b]			Amino acids (MG)	Divergence time estimates (Ma) from different methods[c]						
		Total	MG	SG		MGG_{LS}	MGL_{LS}	SGG_{LS}	SGL_{LS}	SGL_{DT}	SGL_{PL}	Summary
5	Arthropoda vs. Vertebrata	151	120	82	49,644	1,070 ± 101	964 ± 132	942 ± 101	908 ± 132	975 ± 86	994 ± 21	976 ± 97 (786–1166)
6	Cnidaria vs. Bilateria	26	19	17	6,790	1,288 ± 71	1,136 ± 76	1,314 ± 71	1,243 ± 76	1,339 ± 210	1,468 ± 74	1,298 ± 74 (1153–1443)
7	Porifera vs. Eumetazoa	22	19	17	7,090	1,382 ± 151	1,282 ± 122	1,341 ± 151	1,285 ± 122	1,361 ± 211	1,457 ± 54	1,351 ± 120 (1116–1586)
9	Pyrenomycetes vs Plectomycetes	29	18	12	8,952	591 ± 75	559 ± 67	435 ± 75	564 ± 67	654 ± 62	500 ± 19	551 ± 61 (431–671)
10	Candida vs. Saccharomyces	29	21	18	8,918	718 ± 108	714 ± 95	834 ± 108	743 ± 95	725 ± 65	604 ± 14	723 ± 84 (558–888)
11	Hemiascomycetes vs. filamentous Ascomycota	51	33	31	11,683	1,071 ± 125	1,031 ± 103	1,066 ± 125	981 ± 103	915 ± 54	826 ± 16	982 ± 94 (798–1166)
12	Archiascomycetes vs. other Ascomycota	72	56	45	24,348	1,119 ± 83	956 ± 93	1,056 ± 83	994 ± 93	1,011 ± 37	920 ± 12	1,009 ± 73 (866–1152)

13	Basidiomycota vs. Ascomycota	41	27	19	8,504	1,056 ± 165	1,021 ± 112	969 ± 165	927 ± 112	975 ± 48	862 ± 25	968 ± 116 (741–1195)
14	Mucorales/Blastocladiales vs. Basidiomycota/Ascomycota	24	16	15	5,202	1,056 ± 61	1,022 ± 54	921 ± 61	901 ± 54	935 ± 50	845 ± 24	947 ± 51 (847–1047)
15	Fungi vs. animals	188	92	69	31,362	1,594 ± 106	1,492 ± 46	1,511 ± 106	1,449 ± 46	1,435 ± 225	1,594 ± 24	1,513 ± 66 (1384–1642)
17	Mosses vs. vascular plants	51	47	46	4,898	777 ± 103	702 ± 127	677 ± 103	638 ± 127	*1,006* ± 62	742 ± 28	707 ± 98 (515–899)
18	Chlorophytan green algae vs. higher plants	74	63	58	14,333	1,055 ± 76	946 ± 145	921 ± 76	901 ± 145	1,138 ± 45	845 ± 22	968 ± 93 (786–1150)
19	Rhodophyta vs. Chlorophyta+Embryophyta	50	46	43	8,673	1,465 ± 109	1,449 ± 71	1,507 ± 109	1,382 ± 71	1,445 ± 34	1,319 ± 27	1,428 ± 77 (1277–1579)
20	Plants vs. animals	188	143	99	60,274	1,554 ± 67	1,524 ± 53	1,486 ± 67	1,502 ± 53	1,708 ± 283	1,878 ± 26	1,609 ± 60 (1491–1727)
21	Alveolates vs. plants+animals+fungi	76	60	44	21,031	2,086 ± 83	1,807 ± 95	2,011 ± 83	1,903 ± 95	1,972 ± 49	2,057 ± 35	1,973 ± 78 (1820–2126)

22	Euglenozoans vs. plants+animals+fungi	99	80	52	27,759	1,968 ± 56	1,887 ± 68	2,018 ± 56	1,863 ± 68	2,010 ± 61	2,020 ± 35	1,961 ± 57 (1849–2073)
23	Giardia vs. plants+animals+fungi	45	32	28	11,251	2,276 ± 202	2,421 ± 258	2,424 ± 202	2,287 ± 258	2,153 ± 69	2,295 ± 51	2,309 ± 194 (1929–2689)

[a]nodes correspond to the eukaryote phylogenetic tree (Fig. 2). Nodes 1–4 are fossil times and nodes 8 and 16 are phylogenetically constrained (see legend to Fig. 2) and are not shown. b-MG, constant rate proteins used in multigene analyses; SG, proteins used in supergene (concatenated) analyses. c-all error terms, except those for SGLDT, are standard errors of the mean. For SGLDT, the standard deviation is presented, and "credibility intervals" (asymmetric) for that method are presented in Supplemental Table 1 (see 1). Averages of all times and standard errors, excluding one outlier (italics), are shown in the column labeled "Summary," along with the 95% confidence interval.

Hedges et al. BMC Evolutionary Biology 2004 4:2 doi: 10.1186/1471-2148-4-2

Table 2: Estimates of the number of cell types in eukaryotes at different times in past

Group or node[a]	Age (Ma)	Maximum cell types	
		Squared-change parsimony	Linear parsimony
Node 1	310	118	120
Node 2	360	114	120
Node 3	450	104	120
Node 4	520	77	68
Node 9	558	6.1	5
Node 17	717	28.5	26
Node 10	723	3.1	3
Node 14	954	5.7	3
Node 18	956	15.5	9.5
Node 5	974	60	68
Node 13	974	5.7	3
Node 11	985	4.3	3
Node 12	1024	3.7	3
Node 6	1308	33.9	22
Node 7	1355	19.9	16
Node 19	1423	12.7	9.5
Node 8	1450	9.7	3
Node 15	1547	7.3	3
Node 16	1586	6.3	3
Node 20	1624	8.7	8.5
Node 22	1956	7.2	8.5
Node 21	1956	7.2	8.5
Node 23	2291	3.9	2

[a]Nodes correspond to phylogenetic tree of eukaryotes (Fig. 2).

Hedges et al. BMC Evolutionary Biology 2004 4:2 doi: 10.1186/1471-2148-4-2

The use of all available methods for timing protein sequence data (global and local clocks) and different methods of handling the data (multigene and supergene) resulted in remarkably similar estimates of divergence time (Table 1). On average, the six methods differed only

5.5 (4.6–6.4) % from the mean divergence time for a particular node. The resolution here of an animal-fungi relationship also revealed a faster rate of change (on average) in fungi that resulted in slightly younger (~16%) divergence times than reported previously [24]. We attribute the overall consistency among methods to the large size of the data sets and the use of rate tests to eliminate proteins showing substantial rate variation among taxa. It is known that all molecular clock methods, and especially local clock methods, perform best with the largest data sets [14-16], and greater differences are likely to be encountered when a small number of genes are used and when large rate differences are present.

Tests of the Calibrations

We performed a "consistency test" [20] on our major secondary calibration of 964 Ma for the vertebrate-arthropod divergence to determine if it was consistent (reciprocally) with the primary calibration of 310 Ma; in this case, T1 (vertebrate-arthropod divergence) = 310 × (d(vertebrate-arthropod)/d(bird-mammal)) and T2 (bird-mammal divergence) = 964 × (d(bird-mammal)/d(vertebrate-arthropod)). Of 120 rate constant proteins, 118 (98.4%) showed T1 > T2, thus exhibiting high consistency. In the second half of the test, using the supergene matrix of the 82 rate constant proteins, we compared T2 (317 ± 29 Ma) with the primary calibration (310 Ma) and found it to be within one standard error, thus also showing high consistency. The other two secondary calibrations (animal-fungi and animal-plant) also were found to be consistent using the reciprocity test. For animal-fungi, 87/89 (97.8%) rate constant proteins were consistent with the vertebrate-arthropod divergence, and the corresponding T2 value (952 ± 56 Ma) was within one standard error of 964 Ma. For animal-plant, 132/143 (92.3%) rate constant proteins were consistent with the vertebrate-arthropod divergence, and the corresponding T2 value (989 ± 76 Ma) was within one standard error of 964 Ma.

To explore the effect of alternative fossil calibrations, we estimated the vertebrate-arthropod divergence time using our largest data set with expanded taxonomic representation (43 proteins, 19,183 amino acids, 8 taxa) and a diversity of vertebrate and non-vertebrate fossil constraints (lower bounds). The constraints were *Drosophila-Anopheles* (250 Ma), *Homo-Mus* (65 Ma), vertebrate-arthropod (540

Ma), *Saccharomyces-Shizosaccharomyces* (400 Ma) and animal-plant (1200 Ma) [12,25,26]. These constraints are less robust than the bird-mammal calibration (310 Ma), involve smaller numbers of proteins, and probably represent greater underestimates of the true divergence. Nonetheless, the Bayesian (SGL_{DT}) and Penalized likelihood (SGL_{PL}) methods yielded vertebrate-arthropod time estimates of 823 ± 167 and 1289 ± 206 Ma (respectively), still considerably predating the expected time (540 Ma) based on the animal fossil record. Eliminating the two vertebrate fossil constraints resulted in similar time estimates (816 ± 173 and 1285 ± 206 Ma, respectively).

Increase in Cell Types through Time

The maximum cell types of organisms at different time periods are shown in Fig. 3, using data from living organisms and estimates of cell types in common ancestors (Table 2). The origin of life and divergence of archaebacteria and eubacteria were set at 4000 Ma and the origin of eukaryotes at 2700 Ma [27,28], although earlier values for those events would not affect the overall trend, showing a baseline of about 2 cell types in prokaryotes. The results show an increase beginning about 2500 Ma to ~10 cell types at 2000 Ma, and then a second increase from 10–50 between 1500-1000 Ma (Fig. 3).

DISCUSSION

Until the late Proterozoic (~600 Ma), oxygen levels remained low [29], probably limiting the size of eukaryotes, except in photosynthetic algae. However, such algae would not have occurred prior to the origin of plastids (approximately 1600-1500 Ma; Fig. 2) unless they acquired photosynthetic abilities through independent symbiotic events. This would argue against the interpretation of the older (>1600 Ma) fossils of "*Grypania*" as photosynthetic eukaryotic algae [30] and supports their interpretation as colonial prokaryotes [31].

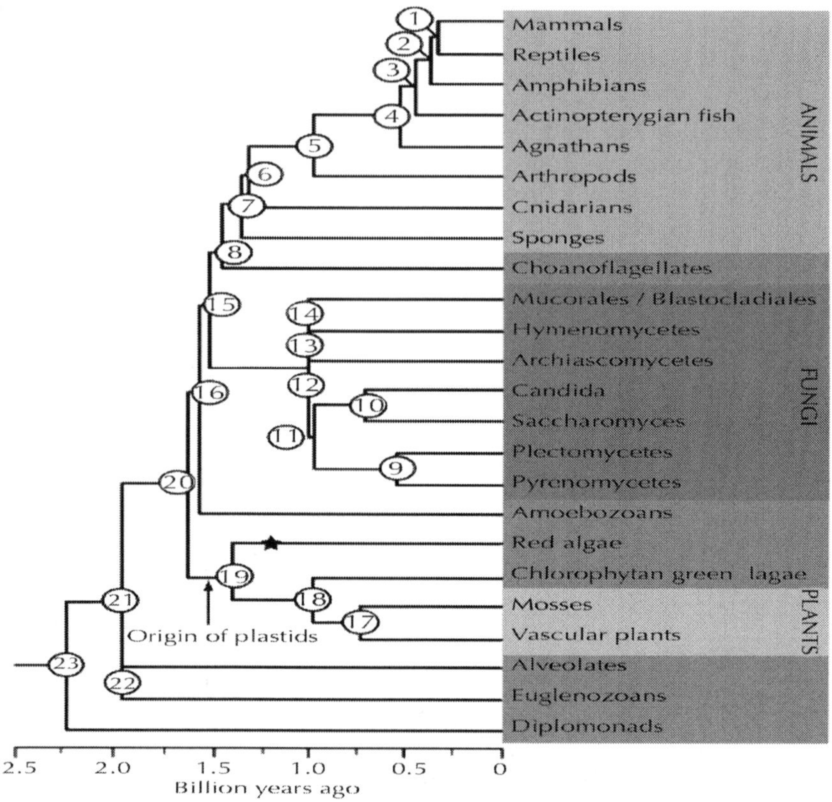

Figure 2: A timescale of eukaryote evolution. The times for each node are taken from the summary times in Table 1, except for nodes 1 (310 Ma), 2 (360 Ma), 3 (450 Ma), and 4 (520 Ma), which are from the fossil record [25]; nodes 8 (1450 Ma) and 16 (1587 Ma) are phylogenetically constrained and are the midpoints between adjacent nodes. Nodes 12–14 were similar in time and therefore shown as a multifurcation at 1000 Ma; likewise, nodes 21–22 are shown as a multifurcation at 1967 Ma. The star indicates the occurrence of red algae in the fossil record at 1200 Ma, the oldest taxonomically identifiable eukaryote [12].

The most frequently used measure of organismal complexity has been the number of cell types [1,2,32]. Other possible measures were not deemed useful (e.g., organism size, genome size) or do not yet have sufficient data available from a diversity of eukaryotes (e.g., number of genes, proteins, transcription factors, introns/exons) for this analysis [32,33]. With a refined timescale of eukaryote evolution it is possible to compare the increase in cell types through time with events in biotic and Earth history (Fig. 3). Although the specific pattern depends on the method of reconstructing character change, some general features are evident. Organisms with more than 2–3 cell types (the maximum in prokaryotes) appeared relatively early (~2000 Ma), soon after the surface environment became oxygenated at 2300 Ma (Great Oxidation Event; [34]). Later, cell types increased again, from 10 to at least 50 on the animal lineage (1500-1000 Ma). By the early Phanerozoic (500 Ma), organisms with more than 50 cell types had evolved. Complexity increased independently in fungi and plants, although at lower absolute levels than in animals.

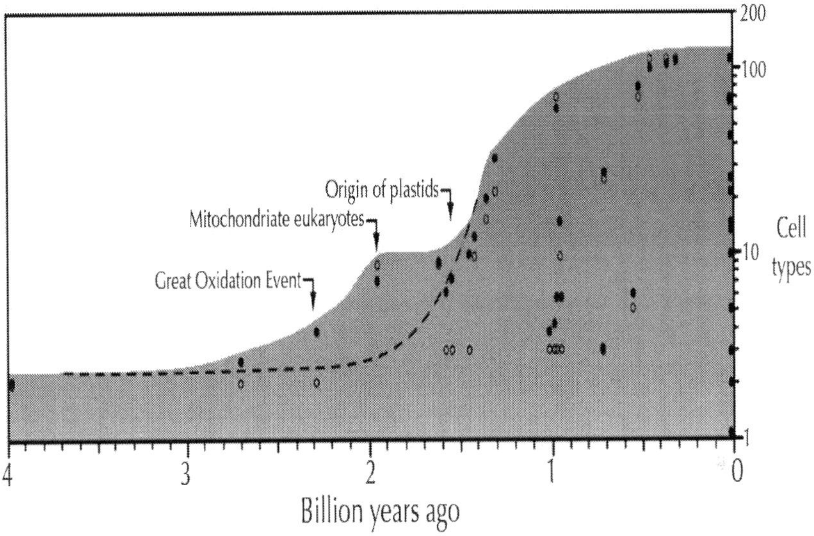

Figure 3: Increase in the maximum number of cell types throughout the history of life. Data points at time zero are from living taxa [1-3,50]; earlier data points were estimated with squared-change parsimony (solid circles) and linear parsimony (hollow circles) [51] using the molecular timetree (Fig. 2). The origin of life and divergence of archaebacteria and eubacteria were set at 4000 Ma and the origin of eukaryotes at 2700 Ma [27,28], although earlier values for those events would not affect the overall trend. We follow McShea [4] in using maximum values at any given time and assuming that decreases do not occur. Dashed line shows an alternate (conservative) interpretation based on uncertainty as to the level of complexity of ancestors of early branching eukaryotes.

There is less confidence in ancestral cell type estimates in the period of initial increase (~2000 Ma) and better support for later estimates (1500-1000 Ma) because of knowledge of gene and structural homology among different groups of animals. For example, it is possible that the last common ancestor of alveolates and higher eukaryotes possessed only one or two cell types rather than the 7–8 predicted in this analysis (Fig. 3; 1973 ± 78 Ma), especially if the rise in complexity was delayed for some reason (e.g., origin of plastids). On the other hand, regardless of when the last common ancestor of protostomes and deuterostomes lived (976 ± 97 Ma in this analysis), there is no doubt that it was a relatively complex (not unicellular) organism with many cell types.

Some early branching eukaryotes (diplomonads) lack mitochondria, although it is debated as to whether they are primitively or secondarily amitochondriate [28]. However, the last common ancestor of mitochondriate eukaryotes, at 1967 ± 65 Ma (Fig. 2), must have possessed a mitochondrion. A molecular clock study of prokaryote and eukaryote genomes [35] arrived at a similar date (1840 ± 200 Ma) for the symbiotic event leading to the mitochondrion, using different data, methods, and approach. This may have been a key event in the rise of complex life, providing eukaryotes with 18 times more energy (over glycolysis alone) for cell signaling and other energy-requiring activities.

CONCLUSIONS

Prior to 2300 Ma, oxygen would not have been widely available for use as an energy source, even if mitochondria existed at that time. Therefore, the initial increase in complexity may have been a response to both energy availability (oxygen) and the ability to extract it (mitochondria). The second and more substantial increase in cell types (1500-1000 Ma) occurred immediately following the acquisition of the plastid (1600-1500 Ma) (Fig. 3), again suggesting a relationship with oxygen. Plastids provided eukaryotes with the ability to generate their own oxygen, benefiting those species (e.g., initially algae and alveolates) directly and their ecosystem partners (e.g., early animals and fungi) indirectly.

METHODS

Data Collection

Nuclear protein sequence data were obtained from the public databases (NCBI Entrez:http://www.ncbi.nlm.nih.gov/entrez/) for all species relevant to each taxonomic comparison, calibration taxa, and outgroups for rate testing (supplemental Table 2; see 2). Initial datasets were screened for orthology using reciprocal BLAST best hits and manual tree building. Additional sequences were also generated from the demosponge, *Microciona prolifera*, for two proteins (enolase and pyruvate kinase). Total messenger RNA was extracted and converted to cDNA pools using reverse transcriptase PCR. Primers were designed from protein sequences available in the public database (enolase forward: 5′ TCCCGYGGKAAYCCMACHGTKGAGGT 3′, reverse: 5′ GGKAGRATCATRAAYTCYTGCATRGC 3′; pyruvate kinase forward: 5′ TTCTCYCAYGGMWCSYACGAGTAYCA 3′, reverse: 5′ CGRAYRAAMGARGCRAASAYCATGTC 3′). Sequences were aligned [36] and regions of ambiguous alignment were removed when necessary. Neighbor-joining trees were constructed (Poisson model) [37] and sequences presumed to be non-orthologous, due to extensive rate variation and evidence of gene duplication, were excluded from further analyses. Short (<100 amino acids) sequences were omitted.

Phylogenetic Analyses

We used a consensus phylogenetic framework based on a diversity of molecular and morphological studies [21, 28]. We also tested six phylogenetic questions with our large protein alignments. The data sets ranged in size from six proteins (3195 amino acids) in the choanoflagellate set to 151 proteins (75,287 amino acids) in the animal-fungi set. All data sets were complete in that they contained all proteins for all species. These concatenated datasets were analyzed using maximum likelihood (JTT + gamma model, quartet puzzling with 1000 steps) [38], minimum evolution (Neighbor-joining, Poisson + gamma model, 2000 bootstraps, complete deletion) [37], and Bayesian Inference (JTT + gamma model, 50,000 generations, 4 chains with starting temp = 0.2) [39]. The shape parameters of the gamma distribution for the different phylogenetic data sets, estimated from the data [40] were: *Giardia* (= 1.12), euglenozoans (= 1.23), alveolates (= 1.18), multiprotist (*Giardia*, euglenozoans, alveolates, plants, animals, fungi) (= 0.93), animal+fungi (= 1.198), plants+red algae (= 0.85), and animals+choanoflagellates (= 0.865).

Calibrations

Times of divergence derived from the fossil record are always underestimates of the true divergence [11,41]. Even the 1200 Ma date for fossil red algae [12] is considered to be an underestimate of the origin of that group because it represents a rare preservation event, hundreds of millions of years older than the next oldest fossil red algae. Therefore, care must be exercised in selecting calibration points or constraints from the fossil record for molecular clock analysis or else they may, in turn, result in considerable underestimates of divergence time [28]. The divergence of the lineages leading to birds and mammals in the fossil record (310 Ma) provides an unusually well-constrained calibration point and permits large numbers of proteins to be used [14]. A more conservative estimate of 288 Ma [42] was used as the lower bound for the mammal-bird divergence in the Bayesian and penalized likelihood analyses; the upper bound was defined by the presence of stem amniotes in the Mid-Late Visean (~345 Ma) [43]. With this primary calibration, we estimated three deeper divergences in the eukaryote tree. In turn, they provided Precambrian calibration points for estimating other divergences. Well-constrained fossil calibration points were otherwise unavailable for the Precambrian. Secondary calibrations minimize the difference between the calibration point and the divergence to be timed, thereby increasing the number of applicable genes and the overall precision of time estimates. For example, genes that show a difference of more than one or two substitutions in a young calibration event (e.g., between two mammals) usually will be evolving too quickly to be alignable or useful for timing deep divergences in eukaryotes. Also, large extrapolations can exaggerate any biases that might exist. Therefore, establishing anchor points or secondary calibrations in the Precambrian permits more genes to be used and reduces the biases caused by large extrapolations.

Divergence Time Estimation

Because the coefficient of variation of time estimates is large for small numbers of genes [14], we used a minimum of 20 genes for each divergence. We chose eighteen divergences among major lineages of eukaryotes, including some analyzed previously [24]. To increase the number of genes available for early branching animals, we sequenced the cDNAs of two genes (enolase and pyruvate kinase) in a poriferan (*Microciona prolifera*) and added those to the assembled data. We subjected all data to global (constant rate) and local (rate variation among lineages) clock methods, including Multigene Global Least Squares (MGG_{LS}) [14], Multigene Local Least Squares (MGL_{LS}) [44], Supergene Global Least Squares (SGG_{LS}) [17], Supergene Local Least Squares (SGL_{LS}), Supergene Local Divtime (SGL_{DT}) [15], and Supergene Local Penalized Likelihood (SGL_{PL}) [16]. The first four (least squares) methods are distance based, SGL_{DT} is a Bayesian method, and SGL_{PL} is a semi-parametric likelihood method. Multigene methods treat each gene separately whereas supergene methods use concatenations of genes [19,41].

All proteins were tested for rate constancy [45,46]; those rejected at the 5% significance level were excluded from timing analyses. Gene-specific and supergene gamma shape parameters ( ) were calculated [40] and used for distance and time estimation [45]. For MGG_{LS}, MGL_{LS}, SGG_{LS}, and SGL_{LS} methods, gene- or supergene-specific rates of sequence change were estimated using linear regression (y-intercept fixed through the origin) from one or more calibrations and applied to the intergroup distance estimates to produce gene- or supergene-specific times. The mode was used as the measure of central tendency in the multigene analyses due to the sensitivity of the mean to extreme values [47]; standard errors of the mode were obtained with bootstrapping (10,000 replications); outliers were trimmed for the supergene data sets.

The SGL_{DT} method was performed using Divtime5b [15]; maximum likelihood branch lengths were calculated under a JTT model using an accompanying program, ESTBRANCHES. The means of the prior distributions ("priors") for the rate parameter and the root time (rt and t, respectively) were calculated for each dataset (see Supplemental Table 1 for parameters). Calibration nodes were constrained using the

95% confidence interval of the secondary calibrations (as discussed previously). Divergence time "posteriors" and their 95% credibility intervals were recorded for each dataset. The SGL_{PL} method was performed in R8S version 1.6 [48] with maximum likelihood branch lengths calculated under a PC+gamma model [40]. A cross-validation procedure [16] was used to obtain the optimal smoothing parameter for each dataset. One hundred bootstrapped datasets were generated to obtain the mean and error on divergence time estimates for each dataset [40,49]. While it is possible to constrain nodes using penalized likelihood, we found that the use of constraints forced the method to overestimate extrapolations and underestimate interpolations (data not shown). For this reason we chose to use fixed calibrations to estimate divergence times with penalized likelihood.

Estimation of Ancestral Numbers of Cell Types

The maximum numbers of cell types in major groups of living organisms were obtained from the literature [1-3,50]: Mammalia (120), Reptilia (120), Amphibia (120), Actinopterygii (120), Arthropoda (69), Agnatha (67), vascular plants (44), mosses (26), Cnidaria (22), Porifera (16), red algae (14), alveolates (14), Pyrenomycetes (9), Hymenomycetes (9), Plectomycetes (9), chlorophytes (5), Saccharomyces (3), Mucorales/Blastocladiales (3), amoebozoans (3), *Candida*(2), Choanoflagellata (2), Euglenozoans (2), diplomonads (2), eubacteria (2), archaebacteria (2), and Archiascomycetes (1). These were used to estimate the maximum number of cell types of common ancestors. This was accomplished with linear and squared change parsimony [51] and the phylogenetic relationships of the groups. Linear and squared change parsimony are preferred over other more complicated methods when all species are extant (as they must be here, for accurate counts of cell types) [52]. Linear parsimony yields more conservative (in this case, lower) estimates than squared change parsimony when a trend is present. For some nodes, linear parsimony yields a range of values; in those cases we followed Webster and Purvis [52] in using the midpoint of the range. The two multifurcations in Fig. 2 were used with squared-change parsimony. Linear parsimony cannot be used with multifurcations and therefore the fungal multifurcation was resolved as (Mucorales/Blastocladiales (Hymenomycetes (Archiascomycetes ((*Candida, Saccharomyces*), (Plectomycetes, Pyrenomycetes))))) and the basal protist multifurcation was resolved as (Diplomonads (Euglenozoans (Alveolates, other eukaryotes))); alternative resolutions did not affect the trend in cell type number.

AUTHORS' CONTRIBUTIONS

SBH directed the research and drafted the manuscript. JEB carried out the bioinformatics research and timing analyses. MLV and JLS assisted in data collection and analysis, and JLS collected new sequences from *Microciona*.

ACKNOWLEDGEMENTS

We thank D. Boone, B. Eidell, J. Hughes, M. Lyons-Weiler, L. Poling, P. Shah, and H. Stone for assistance in the laboratory; J. Hines for artwork; and D. Pisani and J. L. Thorne for discussion. JLS was supported by the Beckman Scholars Program. Other funding was provided by grants to SBH from the National Aeronautics and Space Administration (Astrobiology Institute; NCC2-1057 and NNA04CC06A) and National Science Foundation (DBI-0112670).

REFERENCES

1. Bonner JT: *The evolution of complexity by means of natural selection*. Princeton, New Jersey, Princeton University Press; 1988.
2. Valentine JW, Collins AG, Meyer CP: Morphological complexity increase in metazoans. *Paleobiology* 1994, 20:131-142.
3. Bell G, Mooers AO: Size and complexity among multicellular organisms. *Biological Journal of the Linnean Society* 1997, 60:345-363.
4. McShea DW: The hierarchical structure of organisms: a scale and documentation of a trend in the maximum. *Paleobiology* 2001, 27:405-423.
5. Ward PD, Brownlee D: *Rare Earth*. New York, Copernicus; 2000:333.

6. Wray GA, Levinton JS, Shapiro LH: Molecular evidence for deep Precambrian divergences among metazoan phyla. *Science* 1996, 274:568-573.
7. Gu Xun: Early metazoan divergence was about 830 million years ago. *Journal of Molecular Evolution* 1998, 47:369-371.
8. Wang DY, Kumar S, Hedges SB: Divergence time estimates for the early history of animal phyla and the origin of plants, animals and fungi. *Proceedings of the Royal Society of London Series B-Biological Sciences* 1999, 266:163-171.
9. Runnegar B: A molecular-clock date for the origin of the animal phyla. *Lethaia* 1982, 15:199-205.
10. Ayala FJ, Rzhetsky A, Ayala FJ: Origin of the metazoan phyla: molecular clocks confirm paleontological estimates. *Proceedings of the National Academy of Sciences (U.S.A.)* 1998, 95:606-611.
11. Benton MJ, Ayala FJ: Dating the tree of life. *Science* 2003, 300:1698-1700.
12. Butterfield NJ: Bangiomorpha pubescens n. gen., n. sp.: implications for the evolution of sex, multicellularity, and the Mesoproterozoic/Neoproterozoic radiation of eukaryotes. *Paleobiology* 2000, 26:386-404.
13. Javaux EJ, Knoll AH, Walter MR: Morphological and ecological complexity in early eukaryotic ecosystems. *Nature* 2001, 412:66-69.
14. Kumar S, Hedges SB: A molecular timescale for vertebrate evolution. *Nature* 1998, 392:917-920.
15. Kishino H, Thorne JL, Bruno WJ: Performance of a divergence time estimation method under a probabilistic model of rate evolution. *Molecular Biology and Evolution* 2001, 18:352-361.
16. Sanderson MJ: Estimating absolute rates of molecular evolution and divergence times: a penalized likelihood approach. *Molecular Biology and Evolution* 2002, 19:101-109.
17. Nei M, Xu P, Glazko G: Estimation of divergence times from multiprotein sequences for a few mammalian species and several distantly related organisms. *Proceedings of the National Academy of Sciences (U.S.A.)* 2001, 98:2497-2502.
18. Rodriguez-Trelles F, Tarrio R, Ayala FJ: A methodological bias toward overestimation of molecular evolutionary time scales.

Proceedings of the National Academy of Sciences (U.S.A.) 2002, 99:8112-8115.

19. Hedges SB, Kumar S: Genomic clocks and evolutionary timescales. *Trends in Genetics* 2003, 19:200-206.
20. Shaul S, Graur D: Playing chicken (Gallus gallus): methodological inconsistencies of molecular divergence date estimates due to secondary calibration points. *Gene* 2002, 300:59-61.
21. Baldauf SL: The deep roots of eukaryotes. *Science* 2003, 300:1703-1706. |
22. Sogin ML: History assignment: when was the mitochondrion founded? *Current Opinion in Genetics and Development* 1997, 7:792-799.
23. Stechmann A, Cavalier-Smith T: Rooting the eukaryote tree by using a derived gene fusion. *Science* 2002, 297:89-91.
24. Heckman DS, Geiser DM, Eidell BR, Stauffer RL, Kardos NL, Hedges SB: Molecular evidence for the early colonization of land by fungi and plants. *Science* 2001, 293:1129-1133.
25. Benton MJ: *The Fossil Record 2*. 1st edition. London, Chapman and Hall; 1993:845.
26. Taylor TN, Hass H, Kerp H: The oldest fossil ascomycetes. *Nature* 1999, 399:648.
27. Feng D-F, Cho G, Doolittle RF: Determining divergence times with a protein clock: update and reevaluation. *Proceedings of the National Academy of Sciences (U.S.A.)* 1997, 94:13028-13033.
28. Hedges SB: The origin and evolution of model organisms. *Nature Reviews Genetics* 2002, 3:838-849.
29. Knoll Andrew H., Carroll Sean B.: Early Animal Evolution: Emerging Views from Comparative Biology and Geology. *Science* 1999, 284:2129-2137.
30. Han T-M, Runnegar B: Megascopic eukaryotic algae from the 2.1 billion-year-old Negaunee iron-formation, Michigan. *Science* 1992, 257:232-235.
31. Samuelsson J, Butterfield NJ: Neoproterozoic fossils from the Franklin Mountains, northwestern Canada: stratigraphic and paleobiological implications. *Precambrian Research* 2001, 107:235-251.

32. Carroll SB: Chance and necessity: the evolution of morphological complexity and diversity. *Nature* 2001, 409:1102-1109.
33. Szathmary E, Jordan F, Pal C: Molecular biology and evolution. Can genes explain biological complexity? *Science* 2001, 292:1315-1316.
34. Holland HD: Volcanic gases, black smokers, and the Great Oxidation Event. *Geochimica et Cosmochimica Acta* 2002, 21:3811-3826.
35. Hedges SB, Chen H, Kumar S, Wang D-Y, Thompson AS, Watanabe H: A genomic timescale for the origin of eukaryotes. *BMC Evolutionary Biology* 2001, 1:4.
36. Thompson JD, Higgins DG, Gibson TJ: CLUSTALW: improving the sensitivity of progressive multiple sequence alignment through sequence weighting, position-specific gap penalties and weight matrix choice. *Nucleic Acids Research* 1994, 22:4673-4680.
37. Kumar S, Tamura K, Jakobsen I, Nei M: *MEGA: Molecular Evolutionary Genetics Analysis 2.0.* Tempe, Arizona, Arizona State University; 2000.
38. Strimmer K, vonHaeseler A: Quartet puzzling: A quartet maximum-likelihood method for reconstructing tree topologies. *Molecular Biology and Evolution* 1996, 13:964-969.
39. Huelsenbeck J, Ronquist F: *MrBayes Version 3.0.* Uppsala, Sweden, Evolutionary Biology Centre, Uppsala University; 2003.
40. Yang Z: PAML: a program package for phylogenetic analysis by maximum likelihood. *CABIOS* 1997, 13:555-556.
41. Wray GA: Dating branches on the tree of life using DNA. *Genome Biology* 2001, 3:1-7.
42. Lee MSY: Molecular clock calibrations and metazoan divergence times. *Journal of Molecular Evolution* 1999, 49:385-391.
43. Paton RL, Smithson TR, Clack JA: An amniote-like skeleton from the early Carboniferous of Scotland. *Nature* 1999, 398:508-513.
44. Schubart Christoph D., Diesel Rudolf, Hedges S. Blair: Rapid evolution to terrestrial life in Jamaican crabs. *Nature* 1998, 393:363-365.
45. Kumar S: Phyltest: a program for testing phylogenetic hypotheses, ed. 2.0. In*Institute of Molecular Evolutionary Genetics*. University Park, PA, Pennsylvania State University; 1996.

46. Takezaki Nauko, Rzhetsky Andrey, Nei Masatoshi: Phylogenetic test of the molecular clock and linearized trees. *Molecular Biology and Evolution* 1995, 12:823-833.
47. Hedges SB, Shah P: Comparison of mode estimation methods and application in molecular clock analysis. *BMC Bioinformatics* 2003, 4:31.
48. Sanderson MJ: r8s: inferring absolute rates of molecular evolution and divergence times in the absence of a molecular clock. *Bioinformatics* 2003, 19:301-302.
49. Felsenstein J: *PHYLIP version 3.6*. Seattle, Department of Genetics, University of Washington; 2002.
50. Margulis L, Corliss JO, Melkonian M, Chapman DJ: *Handbook of Protoctista*. Boston, Massachusetts, Jones and Bartlett; 1990:914.
51. Maddison WP, Maddison DR: *MacClade*. Sunderland, MA, Sinauer Associates; 1992.
52. Webster AJ, Purvis A: Testing the accuracy of methods for reconstructing ancestral states of continuous characters. *P Roy Soc Lond B Bio* 2002, 269:143-149.

Chapter 4

Molecular Evolution and Diversification of the Argonaute Family of Proteins in Plants

Ravi K Singh[1], Klaus Gase[2], Ian T Baldwin[2], and Shree P Pandey[1]

[1]Department of Biological Sciences, Indian Institute of Science Education and Research Kolkata, Mohanpur Campus, Mohanpur, Nadia 741246, West Bengal, India

[2]Department of Molecular Ecology, Max Planck Institute for Chemical Ecology, Jena 07745, Germany

ABSTRACT

Background

Argonaute (AGO) proteins form the core of the RNA-induced silencing complex, a central component of the smRNA machinery. Although reported from several plant species, little is known about their evolution. Moreover, these genes have not yet been cloned from the ecological model plant, *Nicotiana attenuata,* in which the smRNA machinery is known to mediate important ecological traits.

Results

Here, we not only identify 11 AGOs in *N. attenuata,* we further annotate 133 genes in 17 plant species, previously not annotated in the Phytozome database, to increase the number of plant AGOs to 263 genes from 37 plant species. We report the phylogenetic classification, expansion, and diversification of AGOs in the plant kingdom, which resulted in the following hypothesis about their evolutionary history: an ancestral AGO underwent duplication events after the divergence of unicellular green algae, giving rise to four major classes with subsequent gains/losses during the radiation of higher plants, resulting in the large number of extant AGOs. Class-specific signatures in the RNA-binding and catalytic domains, which may contribute to the functional diversity of plant AGOs, as well as context-dependent changes in sequence and domain architecture that may have consequences for gene function were found.

Conclusions

Together, the results demonstrate that the evolution of AGOs has been a dynamic process producing the signatures of functional diversification in the smRNA pathways of higher plants.

BACKGROUND

Small-RNA (smRNA)-mediated pathways form a fundamental layer of the transcriptional and post-transcriptional gene regulatory network whose complexity is not fully realized [1]-[4]. The core of this process of RNA interference (RNAi) involves the formation of the RNA-induced silencing complex (RISC) with the help of two major factors. The first factor is the growing class of 18-40 nucleotide (nt) non-coding smRNAs, such as microRNAs (miRNAs), and small-interfering RNAs (siRNAs) [1],[5]. These smRNAs act as sequence specific guides for the second component, the AGOs [4],[6],[7]. AGOs have been implicated as proteins essential in the gene regulatory mechanisms fundamental to developmental and cellular processes such as mRNA stability/degradation, protein synthesis, and genomic integrity [4],[6],[8]. The AGO proteins have characteristically four domains: an N-terminal domain, the PAZ domain, the MID domain and the PIWI domain [4],[9]. The C-terminus of the protein harbors the MID-PIWI lobes. MID-domains have a 'nucleotide specificity loop' that is involved in recognition and binding of the 5'phosphate of smRNAs, whereas the PIWI domains harbor the capacity to slice due to their characteristic catalytic tetrad, 'D-E-D-H/D', at the active site [4],[9],[10]. The 2-nt overhang at the 3' end of miRNAs is recognized by and anchored in the groove of the hydrophilic cleft of the PAZ domain [10],[11]. The N-terminus probably facilitates the separation of smRNA-mRNA duplex as well as may regulate the slicer activity on the target mRNA by interacting with the 3' end of the guide RNA, as recently shown for *Drosophila melanogaster*AGOs [12].

An AGO was originally discovered in forward genetic screens for genes involved in development in*Arabidopsis thaliana* [13]. Yet, little is known about the evolutionary diversification of these proteins across different plant genomes. In Eukaryotes, AGOs are broadly classified into two paralogous families: the AGO family, which have similarities to the founder member, AGO1 of the Arabidopsis, and the PIWI-like proteins, related to *D. melanogaster* 'P-element induced wimpy testis' (PIWI) proteins [4]. While plants have been reported to encode only the AGO-like paralogs, animal genomes harbor representatives of both groups, whereas Amoebozoa are reported to have only PIWI-like genes [4]. A third group of AGOs is specific to *Caenorhabditis elegans*[14]. These findings suggest that both the families have experienced lineage-specific losses [4]. The number of AGO genes varies from 1 (*Schizosaccharomyces pombe*) to 27 (*C. elegans*; [7],[14]); the AGO genes seem to have undergone multiple gene duplication events, but mainly in plant genomes [7]. Plants such as *Chlamydomonas reinhardtii* and *Physcomitrella patens* ('lower plants') contain 4 and 6 members, respectively [15],[16], whereas 'higher plants' such as *Oryza sativa* (OsAGOs) and *A. thaliana* (AtAGOs) contain 18 and 10 members, respectively [1],[2]. In a phylogenetic classification based on protein similarity, 10 AtAGOs were distributed into 3 phylogenetic clades [4],[7], whereas 18 AGO genes of *O. sativa* were divided into 4 clades [7],[17]. However a comprehensive classification of plant AGOs is still missing.

In plants, duplication events may have resulted in functional diversification of AGOs as well as their biochemical activities [7],[18]. For instance, of the 10 AGOs in Arabidopsis, catalytic activities have been demonstrated for only AGO1, AGO2, AGO4, AGO7 and AGO10 [19]-[21]. AtAGO1 and AtAGO10 preferentially bind to smRNAs with a 5′-Uridine (U), whereas AtAGO2, AtAGO4, AtAGO8 and AtAGO9 prefer smRNAs having a 5′-Adenine (A) [22]-[24], while AtAGO5 has greater affinity to 5′-Cytosine (C) containing smRNA [24]. AtAGO10 preferentially binds to smRNAs of 21-nt length, whereas AtAGO4, AtAGO6 and AtAGO9 bind to 24-nt endogenous smRNAs [23],[24]. AtAGO1 binds to miRNAs that are processed by DCL1 and ta-siRNA processed by DCL4 [23],[24]. Furthermore, 82% of smRNAs that associate with AtAGO1 are miRNAs [23], whereas, approximately 11, 2 and 5% of miRNAs are associated with AtAGO2, AtAGO4 and AtAGO5, respectively [23]. AtAGO4 has preferences for miRNAs that are processed by DCL3 [25]. AtAGO4, AtAGO6 and AtAGO9 participate in the RNA-directed DNA methylation pathway [18], whereas AtAGO1 and AtAGO4 play a role in virus resistance [26],[27]. The large number of AGO genes suggests that the smRNA regulatory pathways in plants has undergone substantial diversification and evolution.

Other than in Arabidopsis, AGOs have been reported in other plant species such as rice, maize, and tomato. These genes, however, have yet been identified in the ecological model plant*Nicotiana attenuata* in which the smRNA machinery is known to mediate important ecological traits such as herbivore resistance, competitive ability and UV-B tolerance [28]-[32]. Here, we identify the AGO family of genes in *N. attenuata* (NaAGO), a plant that grows in agricultural primordial niches and is an important model system for the study of plant-herbivore interactions. Further, we investigated the occurrence of AGO proteins in 17 plant species to identify 133 new AGO proteins in plants. Using integrative biology approach (Figure 1) involving molecular phylogenies, consensus sequence comparisons, signature determination, substitution rate estimations and divergence analysis, we propose a model for the evolutionary history of the AGO family of proteins in plants.

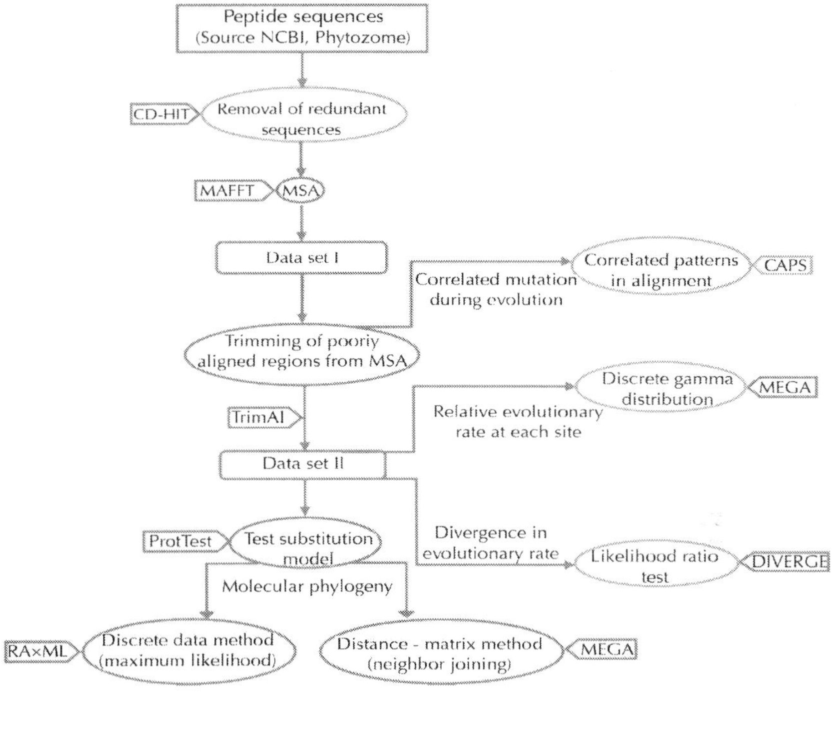

Figure 1: Summary of sequential steps adapted to study evolution of AGOs in plants. A total of 263 AGO sequences from 37 plant species were used in this analysis. Additionally, 5 AGOs from *T. castaneum,* and 1 each from Human (PDB code 4F3T) and *K. polysporus* (PDB code 4F1N) were also used (as out groups) to create the 'AGO dataset I', comprising a total of 270 AGO sequences. The list of AGOs for each species is available in. After MSA and trimming of poorly aligned regions or large gaps, 'AGO dataset II' was generated to contain 270 sequences (rows) and 620 positions.

RESULTS

Data set Assembly and Identification of New AGOs in Plant Genomes

We began with the isolation of 11 unique, full length AGO gene homologs from *N. attenuata*. Putative NaAGOs showed high identity to

8 types of AtAGOs and were thus annotated accordingly as NaAGO1 (identity of >78% to AtAGO1), NaAGO2 (50.55%), NaAGO4 (>74%), NaAGO5 (59.98 %), NaAGO7 (68.95%), NaAGO8 (52.69 %), NaAGO9 (68.04%) and NaAGO10 (80.71%). For NaAGO1, three gene sequences shared >78% peptide identity with AtAGO1 and >87% peptide identity amongst each other; these were thus annotated as NaAGO1a, NaAGO1b and NaAGO1c. Similarly, two gene sequences of the NaAGO4 share 86.86 % peptide identity with each other and >74% identity with AtAGO4; these were named NaAGO4a and NaAGO4b. However, we were not able to identify AtAGO3 and 6 homologs in *N. attenuata*. Further, we mined the sequence data of 17 plant species to identify and similarly annotate 133 full length AGOs. These had not been previously annotated as AGOs. Altogether, 263 protein sequences were used from 37 plant species. Additionally, 5 AGO sequences from*Tribolium castaneum,* 32 AGO sequences (including AGO1 and AGO2) from insects and early branching animals (e.g. sponges, cnideria), and one each of HsAGO2 (PDB code: 4F3T) and KpAGO (PDB code: 4F1N) (for a total of 302 sequences from 66 species;; detailed in methods section) were used as the out-group in this analysis.

Phylogenetic Classification and Evolutionary Expansion of Plant AGOS

- During evolution, AGO genes have formed an expanding family across different lineages [1],[7]. To determine the evolutionary relatedness of plant AGOs, we reconstructed their phylogeny to evaluate their evolutionary patterns (Figure 1). In order to increase the confidence in the root we included 39 non-plant AGO sequences in the phylogenetic analysis. Plant AGO family proved monophyletic and the phylogenetic tree continued to consist of four major classes/clades (Figure 2). Both the Neighbor Joining (NJ) and the Maximum Likelihood (ML) approaches were used to reconstruct the phylogeny of plant AGOs and both produced similar tree topologies and phylogenetic distributions into four classes/clades. Homologs of AGO1 and AGO10 were clustered together (Clade I); similarly homologs of AGOs 2, 3 and 7 formed a clade (Clade III). Likewise, homologs of AGOs 4,

6, 8 and 9 formed the largest cluster (Clade IV), whereas AGO5 homologs formed an independent group (Clade II; Figure 2).

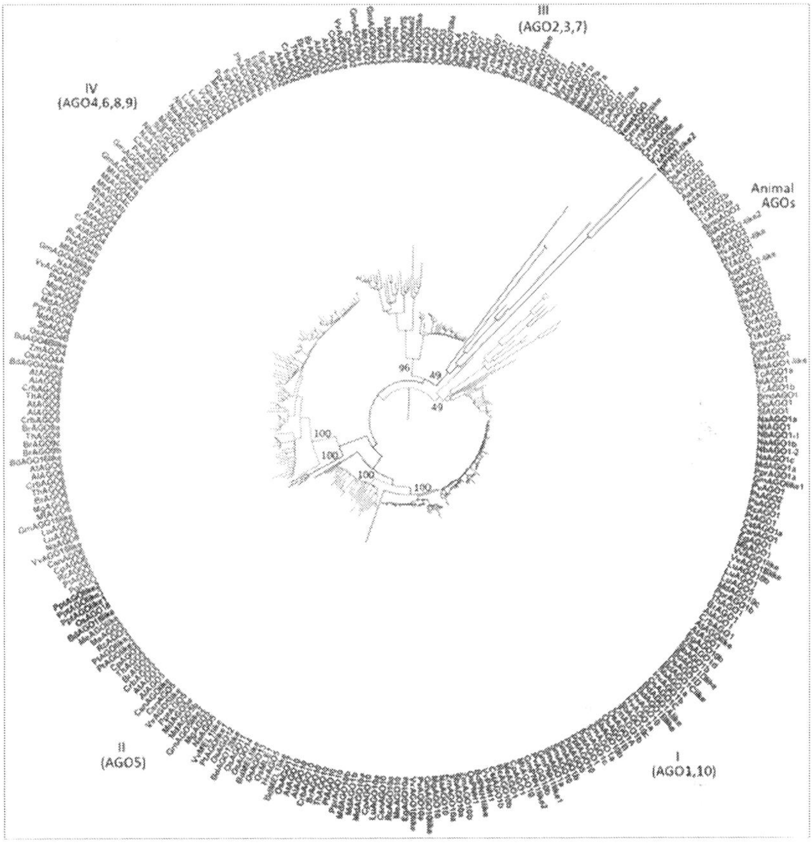

Figure 2: Neighbor joining (NJ) based phylogenetic analysis of AGOs.MEGA 5.2 was used to run the NJ analyses. 39 non-plant AGOs were used to determine the root. Clade robustness was assessed with 100 bootstrap replicates.

From the analysis of AGO gene expansion and loss (detailed in method section), it was observed that AGOs might have undergone between 133-143 duplication and 272-299 loss events. We altered the alignment and alignment processing parameters to test the robustness of our analysis. When L-INSI in MAFFT and 'Automated I' in TrimAl were used, 140 duplication and 299 loss events were obtained; when the parameters were changed to L-INSI (MAFFT) and user defined parameters in TrimAl (detailed in methods section), 133 duplication and 294 loss events were recorded. Similarly, when Auto options were used for both MAFFT and TrimAl, 143 and 294 duplication and loss events were recorded respectively, whereas 137 duplication and 279 loss events were recorded when 'Auto' option in MAFFT and user-defined parameters for TrimAl were used. The reconciliation of species tree and AGO gene family tree (GFT) revealed that the AGO ancestor underwent at least five major duplication events early in its evolution, after the divergence of unicellular green algae, such as, Chlamydomonas and Volvox, but before the divergence of the Bryophytes. This probably gave rise to four distinct phylogenetic clades of AGOs.

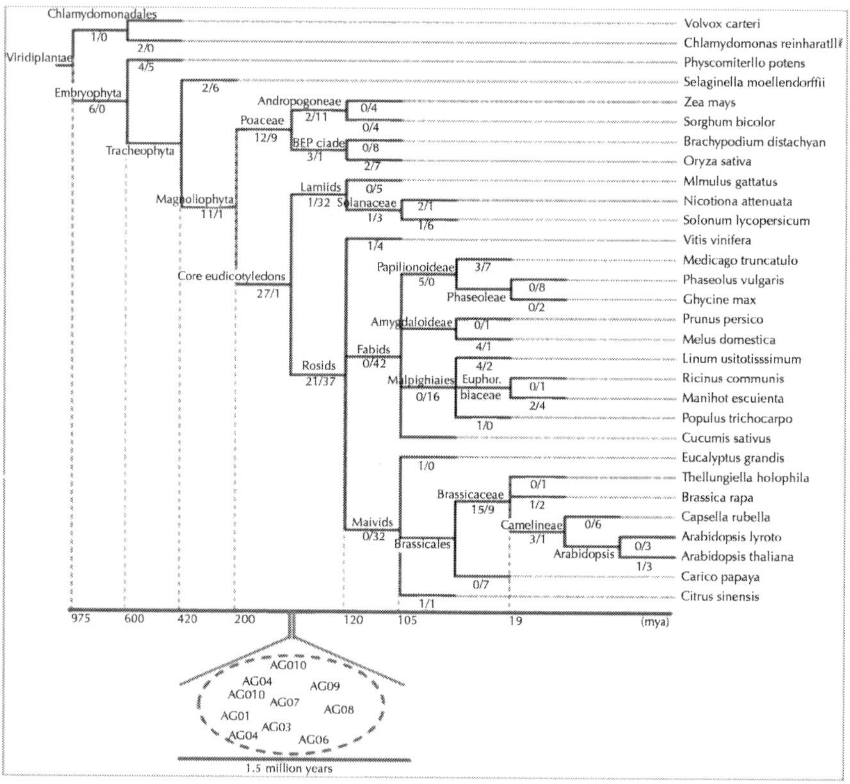

Figure 3: Expansion of AGOs during plant evolution. AGO gene family tree was reconciled with the completely sequenced species tree to identify gain and loss events in each lineage during evolution. The proportions of gains (numerators) versus losses events (denominators) for AGO genes are shown

on each of the branches. Lower panel indicates the tentative time of appearance of different members of the AGO family in plants.

The AGO5 clade may have diverged before the divergence of higher plants, but after the evolution of multicellularity, suggesting a physiological role, possibly different from the ones regulating developmental processes. Reconciliation of AGO GFT with the species tree showed that an ancestral AGO may have undergone >50 rounds of duplications by the time of the dicot-monocot divergence. Thus, diversification and duplication of AGOs could have coincided with the evolution of multicellularity, suggesting the relevance of AGOs and their associated smRNA pathways for developmental and adaptive programs.

The nodes of divergence between dicots and monocots apparent in all four AGO phylogenetic classes indicate that duplications were followed by speciation events. For example, the relatively large number of AGO genes (containing all the four domains) in the Poaceae lineage, such as the 17 in *O. sativa* and the 14 in *Brachypodium distachyon* were noted. These duplication events may have occurred in parallel with events leading to the loss of AGO family members during the evolution of Rosids and Lamiids. Few such losses appeared to have occurred in the Brassicaceae and Solanaceae, for example, in which 10-11 members are found in *A. thaliana* and 11 AGOs in *N. attenuata* respectively. In *N. attenuata*, homologs of AtAGO3 and AtAGO6 might have been lost while AGO1 and AGO4 were duplicated. Duplicated copies of AGO4 are found in other Solanaceae taxa as well, such as in *N. benthamiana*[33] and *Solanum lycopersicum*.

The molecular clock test was performed to gain further insight into the relative timing of duplication and divergence events. This analysis indicates that ancestral AGO gene may have required around 2 million years to duplicate four times after divergence from the unicellular green algae. Clade IV may have been the first to diverge, followed by Clade III, Clade II and Clade I, respectively. It may have taken 0.5 million years for Clade I to evolve that now includes AGO1 and AGO10 homologs, while Clade IV may have required around 1.5 million years to evolve to include AGO4, AGO6, AGO8 and AGO9 homologs; AGO8 and AGO9 as its more recent descendants. Clade III most likely evolved around 1.25 million years and sub-diverged into two clusters, one comprising AGO7 and the other AGO2 and AGO3.

The phylogenetic tree reveals that AGO1 and AGO10 have orthologs in Selaginella and Physcomitrella. Interestingly, we found that of the 6 AGOs in Physcomitrella, the 3 previously unannotated AGO-like genes form a separate cluster (bootstrap value 100%). These AGOs may have diverged from the Clade IV lineage at a time comparable to the duplication of the ancestral AGO, and thus may be orthologs of Class IV AGOs. Furthermore, homologs in unicellular forms, such as Chlamydomonas and Volvox, may have evolved independently from the multicellular lineages. We observed that Chlamydomonas and Volvox AGOs harbor rudimentary forms of the PAZ domain but do not contain a distinct MID domain. These results indicate that AGOs of higher plants are intricate and have substantially diverged from the lower, unicellular forms, potentially to facilitate the complex functions known to be regulated by smRNA pathways.

Variability in Signature Residues of Plant AGOS

Phylogenetic analysis indicates the presence of four clades/classes of AGOs and that these have been evolving differently. In addition, in plants, different AGOs are known to interact with different types of smRNAs (as described in the Background), wherein each residue of the 7nt region of smRNA, 'the seed region', sits in a narrow groove to interact with different residues of the MID-PIWI lobe of AGO proteins [10]. It is hypothesized that the sorting of different species of smRNAs to various AGOs [22],[23] may depend on the conservation of these residues across various AGOs. Such functionally important residues may also be regarded as signatures of specific domains. Therefore, we attempted to define class-wise signature residues for each of the four classes as well as to re-examine the overarching architecture of AGO sequences in plant genomes. The N-terminal domain of AGOs is the most variable domain, whereas, 'R/K-F-Y', 'Y-N-K-K', 'D-E-D-H/D' have been regarded as the signatures of PAZ, MID and PIWI domains, respectively [7]. Upon examining the MSA of all the plant AGOs, we found 55 positions (column score >90) with highly conserved residues. In parallel, we also examined the MSA of plant AGOs in each of the four classes independently to determine class-wise signature residues (Figure 4). We identified 8 sites in the PAZ domains, 12 sites in the MID domains and 15 sites in the PIWI domains that show conservation in the four classes AGOs. In the MID domain, residues 'K', 'Q' and 'C' (alignment position 2485, 2497 and 2498, respectively), thought to directly bind to the 5′-phosphate of smRNAs [10], are conserved in all four classes (Figure 4). Similarly, 'K and 'S' (alignment position 2834 and 2954) of PIWI domain are conserved in all the four classes (Figure 4).

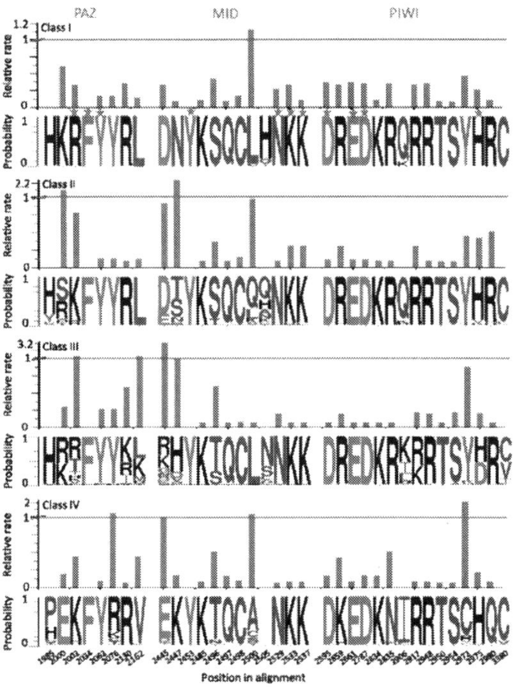

Figure 4: Relative residue bias (probability; lower panel) and relative evolution rate (upper panel) at functionally important positions in the three domains of AGOs in the plant kingdom. Relative frequency of each residue is represented by the height of the corresponding symbol. Height of the bar indicates the relative rate value for respective position. The positions marked with stars (in grey color) are the previously known signature residues.

Results of the MSA indicated that residue 'R', the popularly regarded signature of the PAZ domain ('R-F-Y', alignment positions 2002, 2034 and 2062, respectively), are only conserved in Class I AGOs (AGOs 1 and 10). 'R' has been largely replaced by 'K' (Figure 4) in AGO of Class

II (AGO5) and IV (AGOs 4, 6, 8 and 9), whereas the consensus residue could not be determined for this position (Figure 4) in the PAZ domain of Class III AGOs (AGOs 2, 3 and 7). Further, 'H' at the alignment position 1985 (Figure 4) in the PAZ domain, thought to be important in the recognition of the 3′ ends of smRNAs [10], is conserved only in Classes I-III; conserved residues were not found at this position in the PAZ domain of Class IV genes (Figure 4).

Another residue relevant to the interaction of AGO with the 5′-phosphate of the smRNA in the 'nucleotide specificity loop' of the MID domain is 'T526' (in HsAGO2), which corresponds to alignment position 2447 in plants. Classes I and IV genes harbor a conserved 'N' and 'K' respectively, whereas there is no consensus in Classes II and III at this position. Studies of HsAGO2 [10] suggest that the first oxygen atom of the 5′-phosphate of smRNAs also interacts with side-chain residue of 'R812' in the PIWI domain. Position 2980 corresponds to 'R812', and harbors a conserved 'R' in Classes I-III genes, while in the Class IV genes, PIWI has a conserved 'Q' instead (Figure 4). In the crystal structure of HsAGO2 in a complex with miR-20a, the 2ndnucleotide of smRNA interacts with 'Q548' of the MID domain and 'Q757' of the PIWI domain [10]. These residues correspond to positions 2500 and 2906 in MSA. An 'L' is present at position 2500 in Classes I and III, whereas Classes II and IV are highly variable, with 'Q' and 'A' being over-represented in these two classes respectively (Figure 4). The 5th nucleotide interacts with 'S798' and 'Y804' from the PIWI domain in HsAGO2 [10]. The first corresponding sites in plant AGOs contain 'S' (MSA position 2954) in all four classes, the second site harbors 'Y' (MSA position 2972) in Class I-III whereas 'C' is the over-represented residue in Class IV.

The 'D-E-D-H/D' signature has been associated with the catalytic activity of the PIWI domain [4],[7]. The 'D-E-D-H' signature is apparent in Classes I, II, IV (and half of the Class III) of plant AGOs, whereas the D-E-D-D signature is present in AGO2 and AGO3. In general, most of the functionally important sites of Class-I AGOs are conserved, while the converse seems true for Class-III AGOs (Figure 4).

Since the phylogenetic analysis indicates that the AGOs of unicellular forms such as Chlamydomonas and Volvox are highly divergent and evolved independently of those of the multicellular forms, we further investigated the occurrence of the above-mentioned residue signatures and predicted functionally important sites. We found a high diversity across many important sites. Similarly, the three *Physcomitrella* AGOs also have unique residues compared to AGOs in other lineages.

Such patterns of occurrence of functionally important residues may have consequences for smRNA recruitment, their biochemical activities and the roles of AGOs in diverse physiological processes in both unicellular and multicellular life-forms. Indeed, our homology modeling and RNA docking studies clearly pointed towards differences in seed recognition and catalytic region of the four classes of AGOs.

Evolution of AGO Sequences

We next determined the 'position-by-position' ML-based relative evolutionary rates using a gamma (γ)-distribution based best substitution model. Of the total 620 sites in 'AGO dataset II', 218 sites have a relative rate <1 whereas 69 sites have relative rates >1 in all four classes. Relatively small ML values of γ- shape parameter were observed for Class I, indicating that the majority of sites (405) in Class I AGOs are evolving at slow relative rates. These sites are more frequently found in the MID and PIWI domains (Figure 5). On the other hand, Class III AGOs show a large ML value of the γ- shape parameter, indicating that less number of sites (361 as compared to 405 for example in Class I) are evolving at slow relative rates .

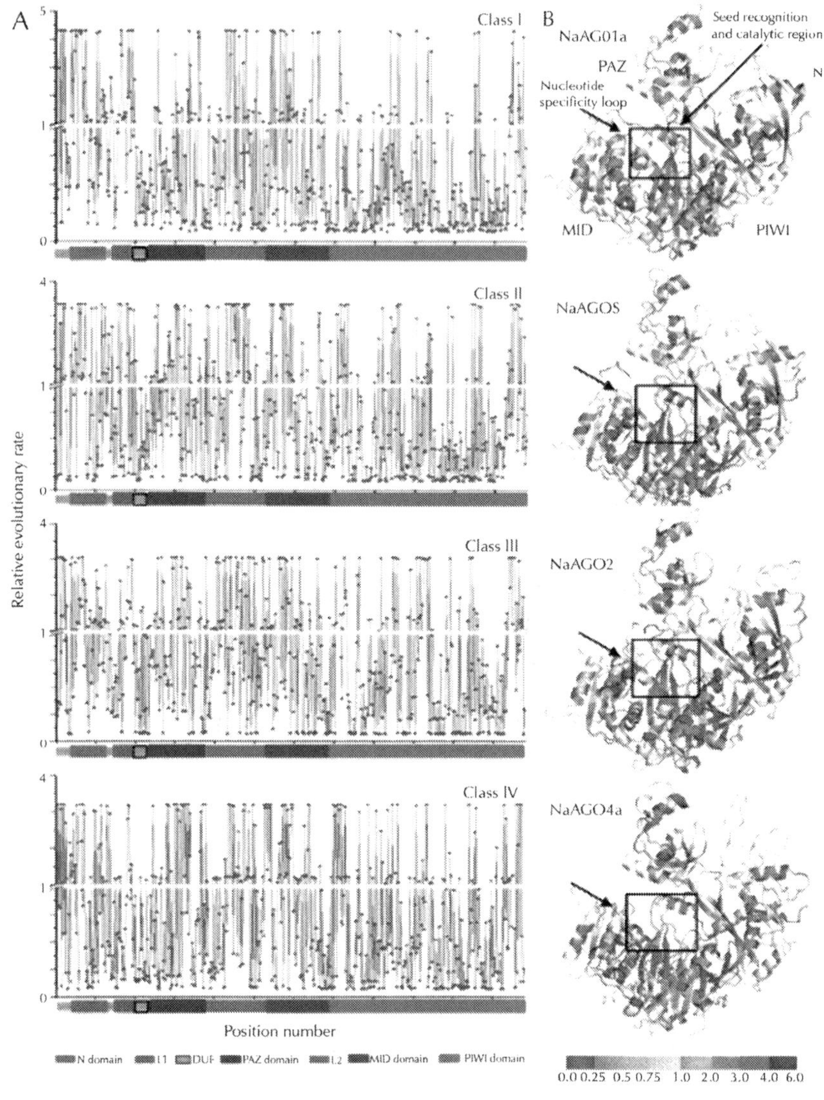

Figure 5: Relative evolutionary rate for each site across four plant AGOs classes. (A) shows site specific relative evolutionary rates of AGOs across classes I-IV. Position-by-position (maximum likelihood) relative evolutionary rates are estimated under the JTT amino acid substitution model. Mean (relative) evolutionary rates are scaled such that the average evolutionary rate across all sites is 1. X-axis represent the positions of residues (620 residues) of the 'AGO dataset II' along the N-terminal, PAZ, MID and PIWI domains in AGO sequence. Y-axis shows the relative evolutionary rate. Sites showing rates <1

are evolving slower than average and those with rates >1 are evolving faster than average. (B) Threaded structures of NaAGO1a, NaAGO5, NaAGO2 and NaAGO4a are modeled as representatives of Classes I-IV respectively, and relative evolutionary rates are mapped on to these structures. Sites with green color represent slow evolving sites (rates <1) and those with red color represent fast evolving sites (rates >1). Different colors in the color bar represent the different rate values.

Residues involved in substrate recognition and catalysis show low relative rates of evolution (Figures 4 and 5), indicating such residues are conserved during the course of evolution. For instance, the 'D-E-D-D/H' signature involved in catalytic activity of the PIWI domain has low relative rates across all the four classes of AGOs. Overall, the seed recognizing MID-PIWI lobe of Classes I and II show a low relative rate (slow evolving; Figure 5). Moreover, other regions putatively involved in seed recognition and the nucleotide specificity loop show a low relative evolutionary rate in Class I AGOs as compared to other classes (Figure 5). For certain sites, substitution of residues along with variability in relative rates was noticed between different classes. For instance, at position 2000, located near the seed recognition pocket and implicated in the 3′ overhang recognition of smRNA [10], substitution of K in Class I to E in Class IV was observed; both the residues are evolving at slow rates (Figure 4). Such changes may explain the capacity of AGO proteins to sort and load smRNAs with specific residues at their termini [23]. On the other hand, it was interesting to note that the N-terminal and the PAZ domains have several sites with high relative rates (fast evolving) across all four classes of AGOs (Figure 5).

These observations suggested the possibility that different classes of AGOs undergo site-specific rate shifts. We performed the likelihood ratio test by calculating the coefficient of Type I (θ_I) divergence and the posterior probability (PP) of a shift in substitution rate. Rejection of the null hypothesis ($\theta_I > 0$) indicates that after duplication, selection constraints may have altered many sites differently in different classes (thus shifts in substitution rates in different classes; θ_I values of 0.2814-0.6509 for pairwise comparisons). Hence, as expected, large variations in site-specific profiles of PP among different classes were observed. Maximum shifts were observed between Classes I and IV. Also, the functional branch lengths (b_F) of Class IV and Class III were nearly two times greater than the branch length of Class I and Class II (p <0.05). Such results point to different evolutionary histories of different classes of AGOs that may have resulted in different structural and functional properties; Class I AGOs may have diverged functionally more than Class IV AGOs.

Context-Dependent Coevolution of Amino Acid Residue

The evolution of protein residues is frequently context-dependent in that substitutions at a given site are affected by local structure, residues at the other sites, and related functions. Such context-dependent substitutions result in co-evolution of amino-acid residues that have implications for protein structure and function. We uncovered coevolving residues in plant AGOs by using Pearson correlation coefficient (r) as implemented in CAPS 2.0 (coevolution analysis using protein sequences) algorithm [34]. Only co-evolving sites with $r \geq 0.5$ were considered significant (Figure 6). Class III AGOs accounted for largest number of coevolving residues (Figure 6A). Strong correlation of $r > 0.9$ was observed between the sites coevolving in the PAZ domain and PIWI domain of Class III AGOs. Four classes of AGOs displayed heterogenous coevolving groups of residues that are of different sizes. In Class III AGOs, PIWI domains displayed the largest number of coevolving residues. In general, the amino acid residue 'R' is the most frequently correlating residue in Class I and II, while residue 'L' is found most frequently correlating in Classes III and IV (Figure 6B). In Class I, 'G' is the second most frequent residue that is significantly correlated mainly to 'G', 'Q', 'R' and 'H'. In Class II, 'G' is again the second most frequent residue that instead significantly correlates to 'V', 'S', 'E', 'K' and 'R' (Figure 6B). In Class III and IV, 'P' is the second most frequent residue that significantly correlates frequently to 'V', 'Q' and 'F', and to 'P', 'G' and 'R' respectively (Figure 6B).

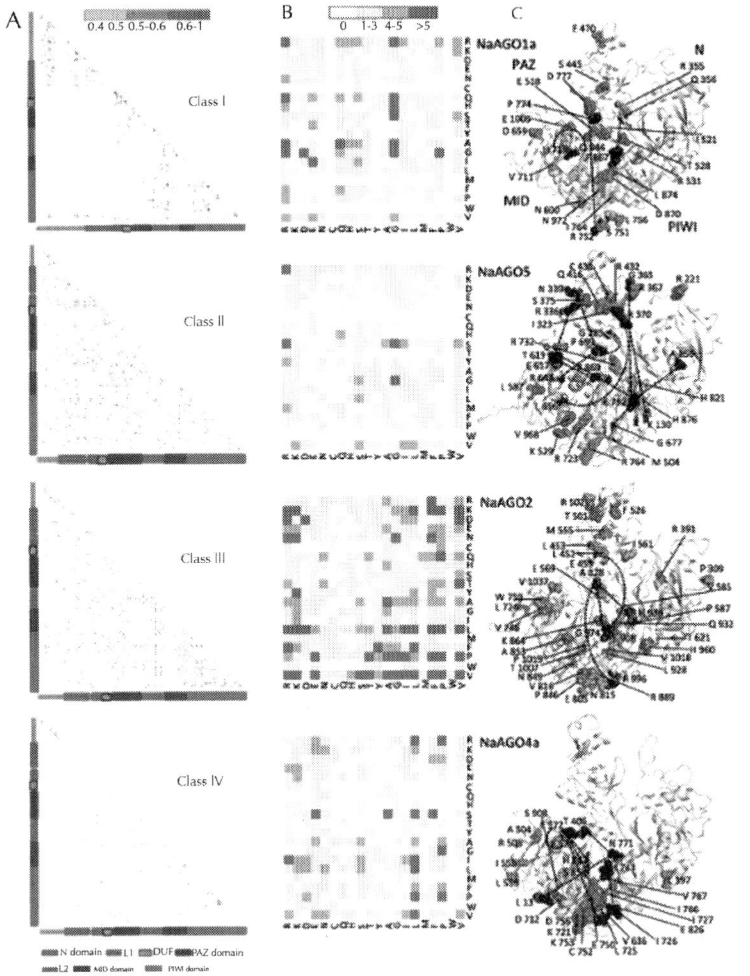

Figure 6: CAPS 2.0 analysis of coevolving sites in plant AGOs. (A) shows heatmaps of coevolving sites in the four classes of plant AGOs. Coevolving pairs showing correlation coefficient of ≥0.5 are plotted. (B) is the color-coded representation of the coevolution frequency matrix of 20 amino acid pairs in NaAGO1a, NaAGO5, NaAGO2 and NaAGO4a, the representatives of four classes respectively. (C) Threaded structures of NaAGO1a, NaAGO5, NaAGO2 and NaAGO4a show the position and arrangement of top coevolving groups in the classes I-IV respectively (orange, green and blue colored; residues in black represent other functionally important coevolving sites as described in Results and Discussion). Residues coded with same color show the correlation with each other in evolutionary context.

Correlation patterns in the context of specific residues at a site in the sequence were observed. For instance, position 2002 (in MSA) in the PAZ domain (may play a role in wedging 14th and 15^{th}nucleotide of loaded RNA duplex [10]), is overrepresented by the residues 'R' in Classes I and III, and 'K' in Classes II and IV AGOs respectively (Figure 4). This 'K' is highly correlated with two other residues in Classes !I and IV (Figure 6C,: Table S5A). On the other hand, position 2002 in Classes I and III does not show any significant correlation coefficient with other residues in the protein. Similarly, the 'H' at position 2505 (Figure 4) in the MID domain of Class I AGO is highly correlated to residue 'Q' at position 2906 in PIWI domain. Residue corresponding to position 2505 (Figure 4) could bind to phosphate of 2nd nucleotide of smRNA, directing the 1st nucleotide into a deep binding pocket at the interface between MID and PIWI domain, whereas Q, corresponding to postion 2906 may coordinate with N2 and N3 on the minor groove side of the G5 base at seed sequence of smRNA [10] . In other classes, where 'H' is replaced, no significant correlation is observed. In HsAGO, 'R' corresponding to 2835 in the PIWI domain stacks between the U9 and U10 of miRNAs to result in a major kink [10]. In Classes I, II and III residues 'R' is conserved at position 2835 (Figure 4) and do not show any correlation with other residues, whereas in Class IV AGOs, this position is overrepresented by 'N' and shows significant correlation to two other residues.

Diverse correlation patterns were observed in the 'nucleotide specificity loop' across the four classes of AGOs (Figure 6). None of the five residues of nucleotide specificity loop of Class I AGO showed any significant correlation. The 5'-end of the smRNAs interact with peptide backbone of the HsAGO residues [10] corresponding to positions 2445 and 2447. 'T', as in HsAGO, was overrepresented only in Class II (AGO5) and correlated to residues in PIWI and PAZ domains (Figure 6C). On the other hand, in Class IV, 'E' (MID domain; position 2445; Figure 4) correlates with 'R' in MID domain and 'V', 'S' and 'H' in PIWI domain. Such class specific coevolving residues may influence the functional diversification of AGOs.

DISCUSSION

Several differences in smRNA processing and mode of action have been noted between plants and animals [1],[3],[35]. In addition, no significant homologies have been found in miRNAs of plants and animals, plants and green algae, or between animals and sponges [1],[3],[36],[37]. This indicates that the smRNA pathways may have evolved independently in the different lineages of life. AGO proteins form the core of the smRNA-mediated regulatory mechanisms and thus are*bonafide* candidates for studying the evolution of smRNA pathways. Here we have reconstructed a comprehensive phylogeny of plant AGO proteins and examined their evolution. Based on this analysis of 302 AGO genes from 66 species, plant AGOs can be divided into four phylogenetic clades/classes. These results suggest that early speciation events separated the AGOs in unicellular and multicellular organisms, wherein AGOs expanded independently to evolve complex domain structures. An ancestral AGO gene may have undergone approximately five duplication events during the time of divergence of green algae and mosses. The AGO family may have further expanded with the emergence of monocot and dicot lineages in plants. Later speciation events may have resulted in species-specific gains or losses of some members.

The smRNA-mediated interaction is biochemically based on the principle of recognition and loading of smRNAs onto the AGOs to form an RNA-protein complex. This complex targets complementary mRNAs and regulates protein synthesis. Diverse pools of smRNAs occur in plant cells that exploit this elegant principle of RNA recognition and cleavage to fine-tune gene expression. Plants produce a large diversity of smRNAs (for instance, miRNAs, tasiRNAs, lsiRNAs, natsiRNAs; [1] that vary in their length (21 nt, 22 nt, 24 nt, and others; [1] and in a preferred base at the 5′ end of smRNAs (e.g. U, A or C; [35]. The particular type of smRNAs that is recruited for executing a particular biochemical RNAi depends on the interaction of the specific smRNA type with an AGO partner [23],[35]. For instance, virus and sense transgene silencing requires the recruitment of 21-22 nt smRNAs onto Class I AGO (AGO1); DNA methylation/chromatin modifications require the association of 24 nt smRNAs onto Class IV proteins (AGOs 4, 6, 9), whereas 21 nt miRNAs recruited onto AGO1 results in mRNA cleavage in plants [23]. The organization of the signatures in RNA-recognition and catalytic domains (PAZ, MID and PIWI) of AGOs play a crucial role in recognizing various forms of smRNAs. Such variations in RNA-interacting domains of AGOs (such as the conservation of R in the signature RFY residues occur only in Class I proteins, the variation of DDH in Class III; conservation of 'N' in the 'nucleotide specificity loop' in the MID domain of only Class I AGOs; Figure 4) may influence substrate recruitment as well as its biochemical fate (such as mRNA degradation, translation inhibition or DNA methylation/chromatin modifications) of the smRNA-target interaction [22],[23],[35].

Here we provide a comprehensive evaluation of plant AGOs: we have not only annotated 133 AGOs in 17 plant species, we have also cloned 11 AGO genes from the wild tobacco plant, *N. attenuata*, a well-studied model system for plant-insect interactions and adaptive plasticity. We have shown earlier that *N. attenuata* harbors at least three functional RdRs [29]-[31],[38] and four DCLs [32]; information on AGO protein in *N. attenuata*had been totally missing. A detailed investigation on the elucidation of function of these 11 NaAGOs could reveal the effector molecules of a unique herbivory-elicited smRNA pathway.

It is evident that many of the functionally important residues are coevolving between the four groups of AGOs. For instance, the residue 'P'. Similarly, position 2800 correlates with only a single polar residue in Classes I, III and IV whereas in Class II, this position correlates with seven residues that may be polar or nonpolar, indicating that the selection pressure is higher at this site in Class II (AGO5) as compared to other classes. Coevolution patterns may form the basis of specialization of AGOs for differential sorting of smRNAs [22],[23]; the 5′end of the incoming smRNA interacts with the peptide backbone of the AGOs. The conformational variability of the residues in the nucleotide specificity loop (Figure 5B and 6C) would make the AGOs selective for specific smRNAs and thus helping them sort different 5′ nucleotides of an incoming RNA [22],[23].

Such specific variations in residues may lead to the functional specialization of AGO proteins. Indeed, specificity in the physiological functions of AGOs has been noticed in Arabidopsis. For instance, AGO4 has been implicated in anti-bacterial defense, whereas AGO1 is implicated in anti-viral response mechanisms [39]. It is noteworthy that although multi-member AGO families have been predicted computationally in several plants, many AGOs in plants (for example in rice and tomato) have not been experimentally confirmed; information on their physiological roles is thus meager. Here we have experimentally determined the full length sequences of 11 *N. attenuata*AGOs. Extensive 'loss-of-function' analyses of these 11 NaAGOs is planned for future investigations.

CONCLUSIONS

The evolution of AGO proteins highly coincides with the evolution of multicellular forms of plants, indicating that smRNAs may have played crucial roles in their development and adaptation. Such an evolutionary pattern of duplication of AGOs also coincides with the evolution of other components of smRNA pathway as recently reported for animal and plant Dicers [40]. Our analysis is consistent with the conclusion that the AGO-mediated recognition of RNA and its regulation is a highly dynamic phenomenon across evolutionary time scales.

METHODS

Dataset Assembly

Data was assembled in three stages: (1) cloning of 11 *N. attenuata* AGOs, (2) mining of novel AGOs and their annotation in plant genomes from public repository, and (3) analysis of already annotated AGO proteins in public databases and sequence repositories. To annotate AGOs, individual AGO transcript and protein sequences (protein sequences wherever not available, were generated by using ORF-finder [http://www.ncbi.nlm.nih.gov/projects/gorf/]) were compared to AtAGOs by using BLAST. While comparing AGO sequences to those of *A. thaliana,* we applied an 'e' value cutoff of $\leq 2e^{-20}$. For isolating *N. attenuata* AGOs, we used a strategy similar to one described earlier [32],[41]. A library made from RNA of *M. sexta* OS-elicited leaf material was sequenced extensively. More than 100 sequenced clones were analyzed to identify full-length NaAGOs. Annotation for each of these NaAGOs was made on the basis of degree of similarity with*A. thaliana* AGOs1-10. Accession numbers of these NaAGOs have been listed. The AGO nucleotide and protein sequence data have been deposited at DDBJ/EMBL/GenBank. Accession numbers of individual NaAGOs are listed. The NaAGO sequences form a part of the Transcriptome

Shotgun Assembly project that has been deposited at DDBJ/EMBL/GenBank under the accession GBGF00000000. The version described in this paper is the first version, GBGF01000000.

NCBI was the primary source of mining of the AGO peptide sequences. A total of 986 unique accession IDs with annotated AGO or AGO like sequences from different plant species available in NCBI (before 30th January, 2012) were retrieved. Further, *A. thaliana* and *O. sativa* AGOs were confirmed from searches of the TAIR and TIGR database, respectively. 'Tomato SBM database' [42] was mined with TBLASTN using default parameters to search for AGO homologs in *Solanum lycopersicum*. Further, the Phytozome database (upto 8th June, 2012; [43]) was mined with BLASTP using default parameter setting to search putative AGO homologs in each completely sequenced species. The CD-HIT program (http://weizhong-lab.ucsd.edu/cdhit_suite/cgi-bin/index.cgi?cmd=cd-hit) [44] with a 90% sequence identity cut-off was used to remove redundant peptide sequences from putative AGOs. Only full-length peptides containing all characteristic domains were retained for further analysis. We annotated each predicted AGO on the basis of their degree of similarity with *A. thaliana* AGOs.

MSA (Multiple sequence alignment) was performed with the help of MAFFT v. 7.130b [45],[46] using the option of 'L-INS-I', a gap-opening penalty of 1.53 and an offset value of 0.123 as well as 'Auto' options. The variable length of many AGOs from different species and numerous poorly aligned regions from the MSA analysis could lead to phylogenetic artifacts. To improve our phylogenetic inferences, we removed the poorly aligned regions [47] to create a trimmed-down version, 'AGO dataset II', by using the program TrimAl v1.3 [48]. To test the robustness of analysis (determine how alignment and alignment processing parameters could affect evolutionary inferences e.g. duplication and loss events), TrimAl was used with 'automated I' option as well as user defined parameters: minimum percentage of position coverage = 10, gap threshold (fraction of positions without gaps in a column) = 0.9, similarity threshold (minimum level of residue similarity within a column) = 0.0. Four parameter combinations were tested to compute duplication/loss events: (i) L-INS-I (MAFFT) + Automated I (TrimAl), (ii) Auto (MAFFT) + Automated I (TrimAl), (iii) L-INS-1 (MAFFT) + User Defined (TrimAl), (iv) Auto (MAFFT) + User Defined (TrimAl). To test the best fit amino acid substitution model and parameter value for 'AGO dataset II' for the tree building analyses, we used ProtTest v 2.4

[49]. The Jones-Taylor-Thornton (JTT) model [50] with an estimated γ-distribution parameter (G) and the proportion of invariant sites (I) was the best fit model according to Akaike Information Criterion (AIC) frame work [51]. This suggests that sequences in the dataset 'Plant AGO dataset II' are closely related with discrete proportion of invariant sites [50].

The MSA of 'AGO dataset I' suggest diverse sequence features in lower groups such as algae. The positions of different domains in the lower plant AGOs were identified by sequence search of the SMART database (http://smart.embl-heidelberg.de/) [52],[53].

Phylogenetic Analysis

Evolutionary relationship among different AGOs was determined using (i) the distance matrix based (NJ) and (ii) the discrete data based methods (ML). NJ analysis was performed with the help of MEGA 5.2 [54] using a JTT model and γ-distributed (G) rate among sites with parameter 1. We tested the homogeneity among lineages to measure the differences in evolutionary patterns for a pair of sequences. Site coverage was kept at 90%; in other words, only sites that have residues in more than 90% sequences were used in the analysis. Clade robustness was assessed with 100 bootstrap replicates. ML analysis was performed by RAxML v 7.2.8 [55]. RAxML analyses were conducted using the PROTCATJTT model with the optimization of substitution rates and site-specific evolutionary rates that are categorized into 25 distinct rate categories for greater computational efficiency [55]. The final tree was evaluated under the γ-distribution of rates and the robustness of clades was assessed with 100 bootstrap replicates.

Expansion of AGOs

To infer the evolutionary history, the AGO gene family tree (GFT) was reconciled with species tree, generated by NCBI Taxonomy Browser (http://www.ncbi.nlm.nih.gov/Taxonomy/CommonTree/wwwcmt.cgi), using NOTUNG program [56], to identify gain and loss events of AGO genes during evolution. Of the 37 species, complete genome sequences of 30 species were available. 248 AGOs from these 30 species were used to analyze gain or loss events. NOTUNG takes a gene family tree, a species tree and a bootstrap threshold as input to generate a gene duplication history as output. The proportion of 'gain' versus 'loss' events is shown on each branch. The numerators and denominators represent the number of gain and loss events respectively of AGO genes during course of evolution (Figure 3). The tentative time of appearance of different members of the AGO family during evolution was calculated using TIME TREE (www.timetree.org) [57],[58].

Molecular Clock

To estimate the divergence time among the AGOs, the molecular clock test was performed by comparing the ML values for a given tree topology with molecular clock constraints to a topology, and without the molecular clock constraints under the JTT (+G + I) model [50] by using MEGA 5.2. Differences in evolutionary rates among sites were modeled using a discrete γ-(G) distribution that allowed for invariant (I) sites to exist. The null hypothesis of equal evolutionary rates throughout the tree was rejected at a 5% significance level.

Structural Modeling

HHpred server (http://toolkit.tuebingen.mpg.de/hhpred/) [59] was used to model the structure of NaAGO1a, NaAGO5, NaAGO2 and NaAGO4a as representatives AGO from classes I-IV respectively. HHpred detects remote protein homology and predicts structure from pairwise comparison of HMM profiles (Hidden Markov models) through various databases search, such as the PDB, SCOP, Pfam, SMART, COGs and CDD. It accepts a single query sequence or a multiple alignment as input and searches through local or global alignments and by scoring secondary structure similarities. HHpred produces pairwise query-template alignments, multiple alignments of the query with a set of templates selected from the search results, as well as 3D structural models that are calculated by the MODELLER from these alignments. Among all the plant AGOs known, the structure of only the MID domain of AtAGO1, 2 and 5 has been resolved to date [22] Human AGO2 (PDB code 4F3T, chain A) was the best template for most of the NaAGOs. ClusPro server (http://cluspro.bu.edu/login.php) [60] was used to dock the 20nt RNA (PDB code 4F3T, chain R) to all four NaAGOs.

Relative Evolutionary Rate and Divergence Analysis

MEGA 5.2 [54] tool was used to estimate the position-by-position (ML) relative evolutionary rate under JTT + G + I amino acid substitution model. The estimate was performed on 90% conservation of sites with 4 rate categories [54]. Individual relative evolutionary rates at all the sites are scaled such that the average evolutionary rate across all positions equals to 1. Hence, positions showing relative rates <1 are more conserved than the average conservation of the sites in the alignment and vice versa. DIVERGE 2.0 [61] was used to identify the sites that show changes in the amino acids substitution rates among different classes (Type I divergence). A trimmed-down version containing 154 sequences of 'Plant AGO dataset II' was used as input to DIVERGE 2.0. A NJ tree for 154 sequences was generated with the help of MEGA 5.2 [54] using a JTT substitution model and clusters for respective classes were selected for pairwise comparison [61]. All paired class comparisons were used to calculate the Coefficient of Type I (θ_I) divergence and the posterior probability of shift in substitution rate for sites. θ_I between different classes provides the statistical evidence for supporting the hypothesis of rate shift between different AGO classes. A tree-like topology for AGO classes was generated that suggests about the functional distance among the classes.

Intramolecular Coevolution in AGOs

CAPS 2.0 (Coevolution analysis using protein sequences;http://bioinf.gen.tcd.ie/~faresm/software/software.html [34]) with default parameters was used to identify co-evolving amino acid site pairs (*e* and *k*) by measuring the correlated evolutionary variation at these sites. Evolutionary variation is measured using divergence time-corrected Blosum values for the transition between two amino acids at a particular site when comparing sequence '*i*' to sequence '*j*' at site '*e*' and '*k*' $(\theta ek)ij$. The time is estimated as the mean number of substitutions per synonymous site between the two sequences being compared. Correlation of the mean variability is measured using the Pearson coefficient. Finally, the significance of the correlation coefficients is estimated by comparing the real correlation coefficients to the distribution of 10,000 randomly sampled correlation coefficients. For the coevolution analyses, a trimmed down version of AGO dataset I′ containing columns without gaps was used (Figure 1). Only coevolving sites showing correlation coefficient of ≥0.5 were considered.

Voucher Specimens

A voucher specimens is deposited at the Herbarium Haussknecht, Jena Germany with following details: *Nicotiana attenuata* Torrey ex Watson "Utah", selfed for 30 generations; cultivation: Max-Planck-Institute for Chemical Ecology, Jena; origin: Washington County, Utah, USA; collected by Ian T. Baldwin, June 3, 1988; GPS coordinates: 37°19′36.26″N 113°57′53.05″W; leg: Tamara Krügel, 18.11.2014, JE. Whole plant including flowers and developing seed capsules was used for nucleic acids isolation.

Deposition of Phylogenetic Trees/Data in Treebase

All the phylogenetic trees and associated matrices have been deposited to the Treebase (http://treebase.org/treebase-web/home.html) with the study number 16716. The Treebase generated URL to access the deposition is http://treebase.org/treebase-web/search/study/summary.html?id=16716 .

AUTHORS' CONTRIBUTIONS

SPP and RKS designed study, RKS, KG and SPP conducted study, ITB and SPP provided resources, RKS, ITB and SPP wrote the MS. All the authors read and approved the Manuscript.

ACKNOWLEDGMENTS

We thank Avinash Sethi and other members of SPP group at IISER-Kolkata for help in computational analyses, and Thomas Hahn and Tamara Krügel, Department of Molecular Ecology, MPI for Chemical

Ecology for help in sequencing and voucher specimen deposition respectively. This work was supported, in part, by the joint MPG-India partner group program of the Max Planck Society and the Indo-German Center for Science and Technology/Department of Science and Technology (India) to SPP and the European Research Council advanced grant ClockworkGreen (No. 293926) to ITB. Funding support of IISER-Kolkata to SPP is also acknowledged.

REFERENCES

1. Axtell MJ: Classification and comparison of small RNAs from plants. *Annu Rev Plant Biol* 2013, 64:137-159.
2. Baulcombe D: RNA silencing in plants.*Nature* 2004, 431(7006):356-363.
3. Eulalio A, Huntzinger E, Izaurralde E: Getting to the root of miRNA-mediated gene silencing. *Cell* 2008, 132(1):9-14.
4. Hutvagner G, Simard MJ: Argonaute proteins: key players in RNA silencing. *Nat Rev Mol Cell Biol* 2008, 9(1):22-32.
5. Mallory AC, Vaucheret H: Functions of microRNAs and related small RNAs in plants. *Nat Genet* 2006, 38(Suppl):S31-S36.
6. Carmell MA, Xuan Z, Zhang MQ, Hannon GJ: The Argonaute family: tentacles that reach into RNAi, developmental control, stem cell maintenance, and tumorigenesis. *Genes Dev* 2002, 16(21):2733-2742.
7. Höck J, Meister G: The Argonaute protein family. *Genome Biol* 2008, 9(2):210.
8. Kuhn CD, Joshua-Tor L: Eukaryotic Argonautes come into focus. *Trends Biochem Sci* 2013, 38(5):263-271.
9. Song JJ, Smith SK, Hannon GJ, Joshua-Tor L: Crystal structure of Argonaute and its implications for RISC slicer activity. *Science* 2004, 305(5689):1434-1437.
10. Elkayam E, Kuhn CD, Tocilj A, Haase AD, Greene EM, Hannon GJ, Joshua-Tor L: The structure of human argonaute-2 in complex with miR-20a. *Cell* 2012, 150(1):100-110.
11. Qi Y, Denli AM, Hannon GJ: Biochemical specialization within Arabidopsis RNA silencing pathways. *Mol Cell* 2005, 19(3):421-428.

12. Hur JK, Zinchenko MK, Djuranovic S, Green R: Regulation of Argonaute slicer activity by guide RNA 3′ end interactions with the N-terminal lobe. *J Biol Chem* 2013, 288(11):7829-7840.
13. Bohmert K, Camus I, Bellini C, Bouchez D, Caboche M, Benning C: AGO1 defines a novel locus of Arabidopsis controlling leaf development. *EMBO J* 1998, 17(1):170-180.
14. Yigit E, Batista PJ, Bei Y, Pang KM, Chen CC, Tolia NH, Joshua Tor L, Mitani S, Simard MJ, Mello CC: Analysis of the C. elegans Argonaute family reveals that distinct Argonautes act sequentially during RNAi. *Cell* 2006, 127(4):747-757.
15. Merchant SS, Prochnik SE, Vallon O, Harris EH, Karpowicz SJ, Witman GB, Terry A, Salamov A, Fritz-Laylin LK, Maréchal-Drouard L, Marshall WF, Qu LH, Nelson DR, Sanderfoot AA, Spalding MH, Kapitonov VV, Ren Q, Ferris P, Lindquist E, Shapiro H, Lucas SM, Grimwood J, Schmutz J, Cardol P, Cerutti H, Chanfreau G, Chen CL, Cognat V, Croft MT, Dent R, et al: The Chlamydomonas genome reveals the evolution of key animal and plant functions.Science 2007, 318(5848):245–250.
16. Rensing SA, Lang D, Zimmer AD, Terry A, Salamov A, Shapiro H, Nishiyama T, Perroud P, Lindquist EA, Kamisugi Y, Tanahashi T, Sakakibara K, Fujita T, Oishi K, Shin-I T, Kuroki Y, Toyoda A, Suzuki Y, Hashimoto S, Yamaguchi K, Sugano S, Kohara Y, Fujiyama A, Anterola A, Aoki S, Ashton N, Barbazuk WB, Barker E, Bennetzen JL, Blankenship R, et al: The Physcomitrella genome reveals evolutionary insights into the conquest of land by plants. Science 2008, 319(5859):64–69.
17. Kapoor M, Arora R, Lama T, Nijhawan A, Khurana JP, Tyagi AK, Kapoor S: Genome-wide identification, organization and phylogenetic analysis of Dicer-like, Argonaute and RNA-dependent RNA Polymerase gene families and their expression analysis during reproductive development and stress in rice. *BMC Genomics* 2008, 9:451.
18. Havecker ER, Wallbridge LM, Hardcastle TJ, Bush MS, Kelly KA, Dunn RM, Schwach F, Doonan JH, Baulcombe DC: The Arabidopsis RNA-directed DNA methylation argonautes functionally diverge based on their expression and interaction with target loci. *Plant Cell* 2010, 22(2):321-334.

19. Baumberger N, Baulcombe DC: Arabidopsis ARGONAUTE1 is an RNA Slicer that selectively recruits microRNAs and short interfering RNAs. *Proc Natl Acad Sci U S A* 2005, 102(33):11928-11933.
20. Fagard M, Boutet S, Morel JB, Bellini C, Vaucheret H: AGO1, QDE-2, and RDE-1 are related proteins required for post-transcriptional gene silencing in plants, quelling in fungi, and RNA interference in animals. *Proc Natl Acad Sci U S A* 2000, 97(21):11650-11654.
21. Rogers K, Chen X: Biogenesis, turnover, and mode of action of plant microRNAs. *Plant Cell* 2013, 25(7):2383-2399.
22. Frank F, Hauver J, Sonenberg N, Nagar B: Arabidopsis Argonaute MID domains use their nucleotide specificity loop to sort small RNAs. *EMBO J* 2012, 31(17):3588-3595.
23. Mi S, Cai T, Hu Y, Chen Y, Hodges E, Ni F, Wu L, Li S, Zhou H, Long C, Chen S, Hannon GJ, Qi Y: Sorting of small RNAs into Arabidopsis argonaute complexes is directed by the 5′ terminal nucleotide.Cell 2008, 133(1):116–127.
24. Takeda A, Iwasaki S, Watanabe T, Utsumi M, Watanabe Y: The mechanism selecting the guide strand from small RNA duplexes is different among argonaute proteins. *Plant Cell Physiol* 2008, 49(4):493-500.
25. Qi Y, He X, Wang XJ, Kohany O, Jurka J, Hannon GJ: Distinct catalytic and non-catalytic roles of ARGONAUTE4 in RNA-directed DNA methylation. *Nature* 2006, 443(7114):1008-1012.
26. Morel JB, Godon C, Mourrain P, Béclin C, Boutet S, Feuerbach F, Proux F, Vaucheret H:Fertile hypomorphic ARGONAUTE (ago1) mutants impaired in post-transcriptional gene silencing and virus resistance. *Plant Cell* 2002, 14(3):629-639.
27. Scholthof HB, Alvarado VY, Vega-Arreguin JC, Ciomperlik J, Odokonyero D, Brosseau C, Jaubert M, Zamora A, Moffett P: Identification of an ARGONAUTE for antiviral RNA silencing in Nicotiana benthamiana. *Plant Physiol* 2011, 156(3):1548-1555.
28. Pandey SP, Shahi P, Gase K, Baldwin IT: Herbivory-induced changes in the small-RNA transcriptome and phytohormone signaling in Nicotiana attenuata. *Proc Natl Acad Sci U S A* 2008, 105(12):4559-4564.

29. Pandey SP, Gaquerel E, Gase K, Baldwin IT: RNA-directed RNA polymerase3 from Nicotiana attenuata is required for competitive growth in natural environments. *Plant Physiol* 2008, 147(3):1212-1224.

30. Pandey SP, Baldwin IT: Silencing RNA-directed RNA polymerase 2 increases the susceptibility of Nicotiana attenuata to UV in the field and in the glasshouse. *Plant J* 2008, 54(5):845-862.

31. Pandey SP, Baldwin IT: RNA-directed RNA polymerase 1 (RdR1) mediates the resistance of Nicotiana attenuata to herbivore attack in nature. *Plant J* 2007, 50(1):40-53.

32. Bozorov TA, Pandey SP, Dinh ST, Kim SG, Heinrich M, Gase K, Baldwin IT: DICER-like proteins and their role in plant-herbivore interactions in Nicotiana attenuata. *J Integr Plant Biol* 2012, 54(3):189-206.

33. Jones L, Keining T, Eamens A, Vaistij FE: Virus-induced gene silencing of argonaute genes in Nicotiana benthamiana demonstrates that extensive systemic silencing requires Argonaute1-like and Argonaute4-like genes. *Plant Physiol* 2006, 141(2):598-606.

34. Fares MA, McNally D: CAPS: coevolution analysis using protein sequences. *Bioinformatics* 2006, 22(22):2821-2822.

35. Khvorova A, Reynolds A, Jayasena SD: Functional siRNAs and miRNAs exhibit strand bias. *Cell* 2003, 115(2):209-216.

36. Bartel B: MicroRNAs directing siRNA biogenesis. *Nat Struct Mol Biol* 2005, 12(7):569-571.

37. Brodersen P, Sakvarelidze-Achard L, Bruun-Rasmussen M, Dunoyer P, Yamamoto YY, Sieburth L, Voinnet O: Widespread translational inhibition by plant miRNAs and siRNAs. *Science* 2008, 320(5880):1185-1190.

38. Pandey SP, Moturu TR, Pandey P: Roles of small RNAs in regulation of signaling and adaptive responses in plants. *Recent Trends Gene Express* 2013, 5:107-132.

39. Ruiz-Ferrer V, Voinnet O: Roles of plant small RNAs in biotic stress responses. *Annu Rev Plant Biol* 2009, 60:485-510.

40. Mukherjee K, Campos H, Kolaczkowski B: Evolution of animal and plant dicers: early parallel duplications and recurrent adaptation of antiviral RNA binding in plants. *Mol Biol Evol* 2013, 30(3):627-641.

41. Halitschke R, Baldwin IT: Antisense LOX expression increases herbivore performance by decreasing defense responses and inhibiting growth-related transcriptional reorganization in Nicotiana attenuata. *Plant J* 2003, 36(6):794-807.

42. Shirasawa K, Hirakawa H: DNA marker applications to molecular genetics and genomics in tomato. *Breed Sci* 2013, 63(1):21-30.

43. Goodstein DM, Shu S, Howson R, Neupane R, Hayes RD, Fazo J, Mitros T, Dirks W, Hellsten U, Putnam N, Rokhsar DS: Phytozome: a comparative platform for green plant genomics.Nucleic Acids Res 2012, 40(Database issue):D1178–D1186.

44. Huang Y, Niu B, Gao Y, Fu L, Li W: CD-HIT Suite: a web server for clustering and comparing biological sequences. *Bioinformatics* 2010, 26(5):680-682.

45. Katoh K, Misawa K, Kuma K, Miyata T: MAFFT: a novel method for rapid multiple sequence alignment based on fast Fourier transform. *Nucleic Acids Res* 2002, 30(14):3059-3066.

46. Katoh K, Standley DM: MAFFT multiple sequence alignment software version 7: improvements in performance and usability. *Mol Biol Evol* 2013, 30(4):772-780.

47. Talavera G, Castresana J: Improvement of phylogenies after removing divergent and ambiguously aligned blocks from protein sequence alignments. *Syst Biol* 2007, 56(4):564-577.

48. Capella-Gutiérrez S, Silla-Martínez JM, Gabaldón T: trimAl: a tool for automated alignment trimming in large-scale phylogenetic analyses. *Bioinformatics* 2009, 25(15):1972-1973.

49. Abascal F, Zardoya R, Posada D: ProtTest: selection of best-fit models of protein evolution. *Bioinformatics* 2005, 21(9):2104-2105.

50. Jones DT, Taylor WR, Thornton JM: The rapid generation of mutation data matrices from protein sequences.*Computer applications in the biosciences: CABIOS* 1992, 8(3):275-282.

51. Posada D, Buckley TR: Model selection and model averaging in phylogenetics: advantages of akaike information criterion and bayesian approaches over likelihood ratio tests. *Syst Biol* 2004, 53(5):793-808.

52. Letunic I, Doerks T, Bork P: SMART 7: recent updates to the protein domain annotation resource. *Nucleic Acids Res* 2012, 40(Database issue):D302-D305.

53. Ponting CP, Schultz J, Milpetz F, Bork P: SMART: identification and annotation of domains from signalling and extracellular protein sequences. *Nucleic Acids Res* 1999, 27(1):229-232.
54. Tamura K, Peterson D, Peterson N, Stecher G, Nei M, Kumar S: MEGA5: molecular evolutionary genetics analysis using maximum likelihood, evolutionary distance, and maximum parsimony methods. *Mol Biol Evol* 2011, 28(10):2731-2739.
55. Stamatakis A: RAxML-VI-HPC: maximum likelihood-based phylogenetic analyses with thousands of taxa and mixed models. *Bioinformatics* 2006, 22(21):2688-2690.
56. Chen K, Durand D, Farach-Colton M: NOTUNG: a program for dating gene duplications and optimizing gene family trees. *J Comput Biol* 2000, 7(3–4):429-447.
57. Hedges SB, Dudley J, Kumar S: TimeTree: a public knowledge-base of divergence times among organisms. *Bioinformatics* 2006, 22(23):2971-2972.
58. Kumar S, Hedges SB: TimeTree2: species divergence times on the iPhone. *Bioinformatics* 2011, 27(14):2023-2024.
59. Soding J, Biegert A, Lupas AN: The HHpred interactive server for protein homology detection and structure prediction.*Nucleic Acids Res* 2005, 33(Web Server issue):W244-W248.
60. Comeau SR, Gatchell DW, Vajda S, Camacho CJ: ClusPro: a fully automated algorithm for protein-protein docking. *Nucleic Acids Res* 2004, 32(Web Server issue):W96-W99.
61. Gu X, Vander Velden K: DIVERGE: phylogeny-based analysis for functional-structural divergence of a protein family. *Bioinformatics* 2002, 18(3):500-501.

Chapter 5

Molecular Mechanisms of Ultraviolet Radiation-Induced DNA Damage and Repair

Rajesh P. Rastogi[1], Richa[1], Ashok Kumar[2], Madhu B. Tyagi[3], and Rajeshwar P. Sinha[1]

[1]Laboratory of Photobiology and Molecular Microbiology, Centre of Advanced Study in Botany, Banaras Hindu University, Varanasi 221005, India

[2]School of Biotechnology, Banaras Hindu University, Varanasi 221005, India

[3]Mahila Maha Vidyalaya, Banaras Hindu University, Varanasi 221005, India

ABSTRACT

DNA is one of the prime molecules, and its stability is of utmost importance for proper functioning and existence of all living systems. Genotoxic chemicals and radiations exert adverse effects on genome stability. Ultraviolet radiation (UVR) (mainly UV-B: 280–315 nm) is one of the powerful agents that can alter the normal state of life by inducing a variety of mutagenic and cytotoxic DNA lesions such as cyclobutane-pyrimidine dimers (CPDs), 6-4 photoproducts (6-4PPs), and their Dewar valence isomers as well as DNA strand breaks by interfering the genome integrity. To counteract these lesions, organisms have developed a number of highly conserved repair mechanisms such as photoreactivation, base excision repair (BER), nucleotide excision repair (NER), and mismatch repair (MMR). Additionally, double-strand break repair (by homologous recombination and nonhomologous end joining), SOS response, cell-cycle checkpoints, and programmed cell death (apoptosis) are also operative in various organisms with the expense of specific gene products. This review deals with UV-induced alterations in DNA and its maintenance by various repair mechanisms.

INTRODUCTION

The stratospheric ozone layer is continuously depleting due to the release of atmospheric pollutants such as chlorofluorocarbons (CFCs), chlorocarbons (CCs), and organo-bromides (OBs). Consequently there is an increase in the incidence of UV radiation (UVR) on the Earth's surface [1] which is one of the most effective and carcinogenic exogenous agents that can interact with DNA and alter the genome integrity and may affect the normal life processes of all organisms ranging from prokaryotes to mammals [2–10]. However, wide variations in tolerance to UV-B among species and taxonomic groups have been reported. Moreover, ozone depletion followed by increased UV exposure has been predicted to continue throughout most of this century [11]. In all the groups of UVR (i.e., UV-A: 315–400 nm; UV-B: 280–315 nm; UV-C: <280 nm) UV-B radiation produces adverse effects on diverse habitats, even though most of the extraterrestrial UV-B is absorbed by the stratospheric ozone [12]. UV-A radiation has a poor efficiency in inducing DNA damage, because it is not absorbed by native DNA. UV A and visible light energy (up to 670–700 nm) are able to generate singlet oxygen ($^1O_2$) that can damage DNA via indirect photosensitizing reactions [13]. UV-C radiation is quantitatively absorbed by oxygen and ozone in the Earth's atmosphere, hence does not show much harmful effects on biota. Solar UV radiation is responsible for a wide range of biological effects including alteration in the structure of protein, DNA, and many other biologically important molecules, chronic depression of key physiological processes, and acute physiological stress leading to either reduction in growth and cell division, pigment bleaching, N_2 fixation, energy production, or photoinhibition of photosynthesis in several organisms [3, 9, 10]. It has been documented that UV-B severely affects survival, fecundity, and sex-ratio in several intertidal copepods [14]. One of the most prominent targets of solar UV-radiation is cellular DNA, which absorbs UV-B radiation and causes adverse effects on living systems such as bacteria [15, 16], cyanobacteria [17], phytoplankton [18], macroalgae [19], plants [20], animals, and humans [21–23].

Although UV-B radiation has less than 1% of total solar energy, it is a highly active component of the solar radiation that brings about chemical modification in DNA and changes its molecular structure by the formation of dimers. Certain UV-absorbing pigments are produced by a number of organisms as a first line of defense; however, they are unable to avoid UV-radiation completely from reaching DNA in superficial tissue [28–32]. Certain enzymes, such as superoxide dismutase (SOD), catalase (CAT), peroxidase (POD), and scavengers such as vitamin C, B, and E, cysteine, and glutathione play an additional role in defense against UV radiation [33]. However, as a second line of defense several organisms have developed a number of specific and highly conserved repair mechanisms such as photoreactivation, excision repair, mismatch repair (MMR), double strand break (DSB) repair and certain other mechanisms like damage tolerance (dimer bypass), SOS (save our soul) response, checkpoint activation, and programmed cell death (PCD) or apoptosis (Figure 1) that efficiently remove DNA lesions ensuring the genomic integrity [22]. Plants are unique in the obligatory nature of their exposure to UVR; it is also conceivable that they may also have evolved certain efficient repair mechanisms for the elimination of UV-induced DNA damage. However, a number of questions concerning the basic phenomena of the DNA repair in plants remain to be elucidated. In the following, we discuss the molecular mechanisms of UV-induced DNA damage and repair mechanism (s) operative in various organisms.

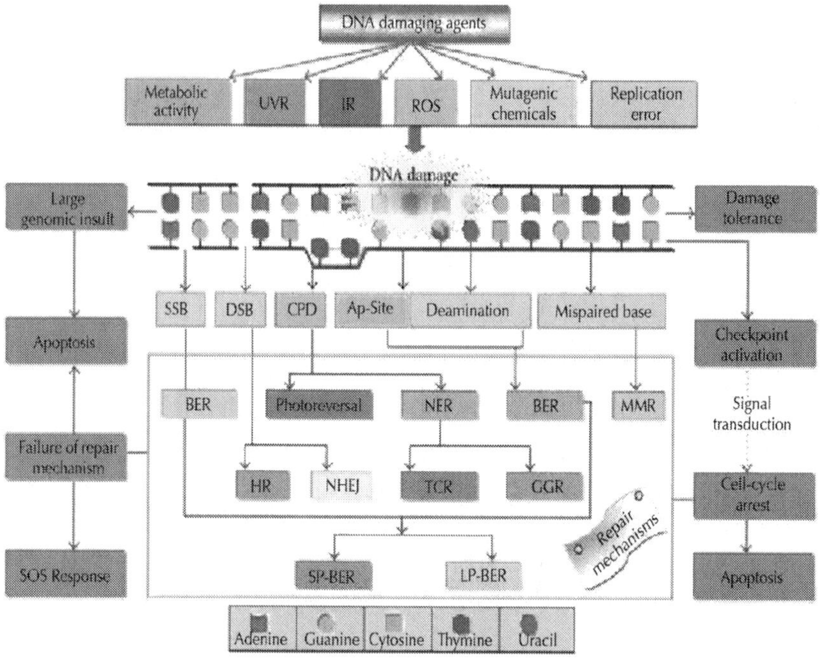

Figure 1: DNA damage and maintenance. Genomic lesions produced by various DNA damaging agents trigger several specific repair machinery to conserve the genomic integrity. In case of severe damage and/or failure of repair mechanisms, cells undergo apoptosis or induce a complex series of phenotypic changes, that is, SOS response. Sometimes the potentiality of lesions in the genome is mitigated by a phenomenon known as damage tolerance, during which DNA lesions are recognized by certain repair machinery, allowing the cells to undergo normal replication and gene expression. The cellular response to DNA damage may activate cell-cycle checkpoint by means of a network of signaling pathway that gives the cell extra time to repair the genomic lesions or may induce cell suicide response/programmed cell death (PCD).

UV-INDUCED DNA DAMAGE

Induction of DNA damage by solar UVR is a key event that drastically influences the normal life processes of all organisms. A number of endogenous factors such as free radicals [34] generated during metabolic processes as well as exogenous factors such as UV or ionizing radiations [35] are known to interfere with genome integrity. DNA damage results in (i) misincorporation of bases during replication process, (ii) hydrolytic damage, which results in deamination of bases, depurination, and depyrimidination [36], (iii) oxidative damage, caused by direct interaction of ionizing radiations (IR) with the DNA molecules, as well as mediated by UV radiation-induced free radicals or reactive oxygen species [37, 38], and (iv) alkylating agents that may result in modified bases [36, 39]. The hydrolytic deamination (loss of an amino group) can directly convert one base to another; for example, deamination of cytosine results in uracil and at much lower frequency adenine to hypoxanthine. In depurination/depyrimidination, there are complete removals of purine/pyrimidine bases, leaving the deoxyribose sugar depurinated/depyrimidinated,that may cause breakage in the DNA backbone. The exposure of UVR, IR, and certain genotoxic chemicals may result in single as well as double DNA strand breaks. Among different types of damages, DNA double strand breaks (DSBs) are the most deleterious, since they affect both strands of DNA and can lead to the loss of genetic material. At high concentrations oxygen-free radicals or, more frequently, reactive oxygen species (ROS) can induce damage to cell structure, lipids, proteins as well as DNA and results in oxidative stress which has been implicated in a number of human diseases [40]. The hydroxyl radicals (OH°) can damage all components of DNA molecules such as purine and pyrimidine bases and also the deoxyribose backbone, inhibiting the normal functions of the cell [37, 38].

UV-INDUCED PYRIMIDINE PHOTOPRODUCTS

UV-B radiation is one of the most important energetic solar components that may lead to the formation of three major classes of DNA lesions, such as cyclobutane pyrimidine dimers (CPDs), pyrimidine 6-4 pyrimidone photoproducts (6-4PPs), and their Dewar isomers (Figure 2) [5, 22, 23, 41–43]. CPDs correspond to the formation of a four-member ring structure involving C5 and C6 of both neighboring bases whereas 6-4PPs are formed by a noncyclic bond between C6 (of the 5′-end) and C4 (of the 3′-end) of the involved pyrimidines via spontaneous rearrangement of the oxetane (when the 3′-end is thymine) (Figure 3(a)) or azetidine (when the 3′-end is cytosine) (Figure 3(b)) intermediates [5, 44]. The 6-4PPs are eagerly converted into their Dewar valence isomers upon exposure to UV-B or UV-A radiation that may further undergo reversion to the 6-4PPs upon exposure to short-wavelength UV radiation [45]. Two adjacent cytosines are considered as mutation hotspots of UVB and UVC radiations [42]. It has been found that T-T and T-C sequences are more photoreactive than C-T and C-C sequences [46]. The diastereoisomers of pyrimidine dimers (Figure 4) can be observed in free solution that differ in the orientation of the two pyrimidine rings relative to the cyclobutane ring, and on the relative orientations of the C5–C6 bonds in each pyrimidine base [44]. It has been demonstrated that the main photoproducts are cis-syn-configured CPD lesions, while trans-syn-configured CPD lesions are formed in much less quantity [47]. In double stranded B-DNA, where the dimer entails two adjoining pyrimidine bases on the same DNA strand, only the syn isomers can be generated, whereas the cis isomer is preferred over the trans isomer to a great extent [42]. The incidence of trans-synisomer in single-stranded or denatured DNA is more common because of the increased flexibility of the DNA backbone. A few CPD lesions (i.e., cis-syn or trans-anti isomers) can also be detected in aqueous solutions by UV-C irradiation [48]. The formation of "spore photoproduct" has been detected in UV-irradiated bacterial spores by the addition of methyl group of one thymine residue to the C5 position of an adjacent thymine. In most cellular environments, there is no much significance of this photoproduct, since it requires anhydrous conditions for its formation [49].

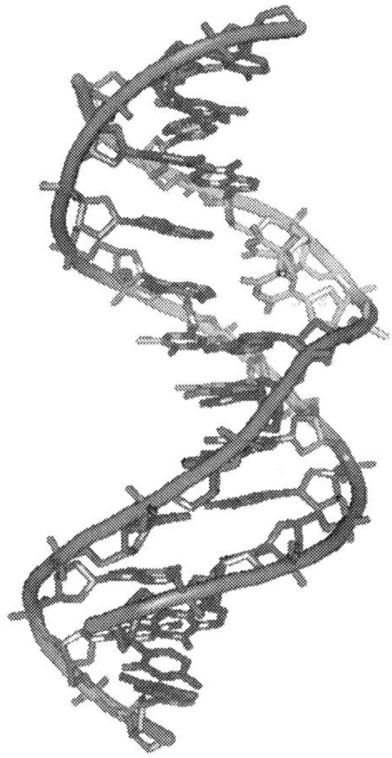

(a)

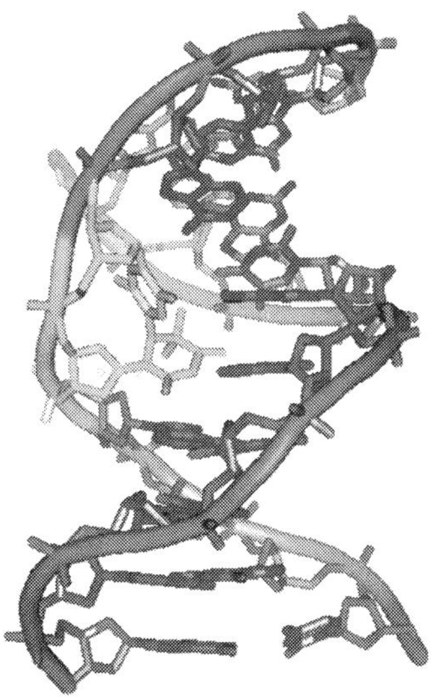

(b)

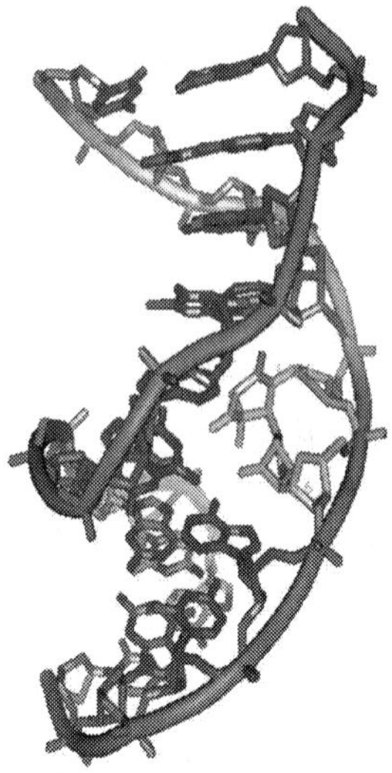

(c)

Figure 2: Structures of DNA duplexes showing the presence of lesions (in green) such as CPD (a), 6-4PP (b), and 6-4 Dewar dimer (c). Hydrogen atoms are not shown, prepared from PDB entries 1TTD [24], 1CFL [25], and 1QKG [26] using PyMol. (version 1.1r1) [27].

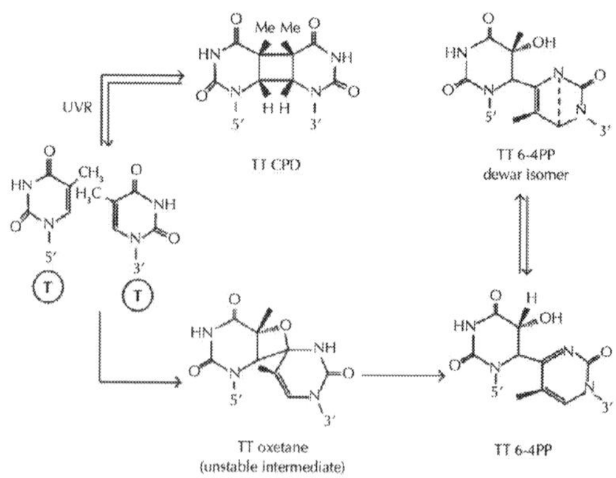

(a)

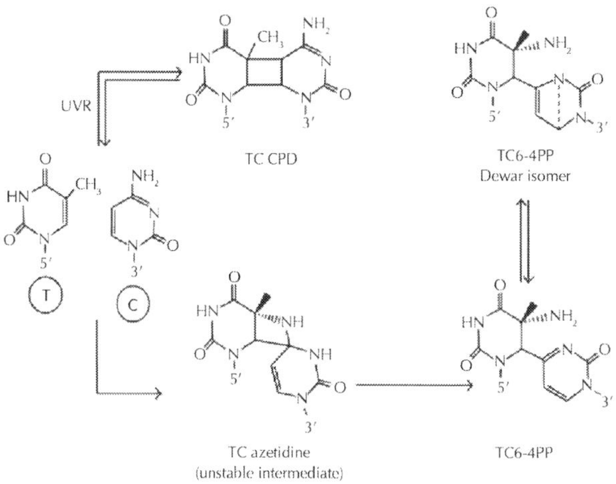

(b)

Figure 3: Pathway of UVR-induced T-T (a) and T-C (b) CPD, 6-4PPs, and their Dewar isomers.

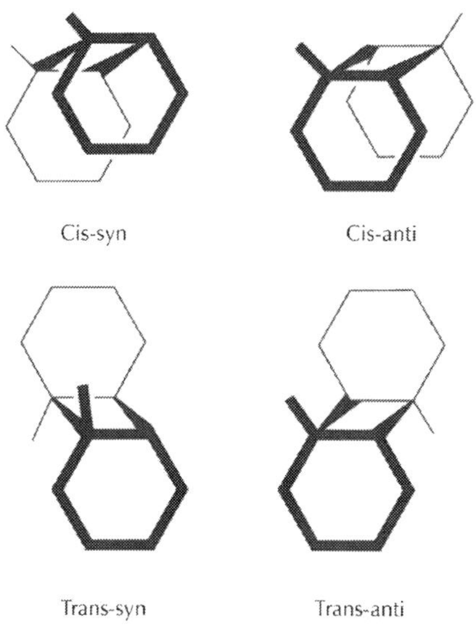

Figure 4: Possible diastereoisomers of pyrimidine T<>T dimer.

The base damage by UVR is determined by the flexibility of the DNA strand as well as nature and position of the base. CPDs are formed at higher quantity by cycloaddition reaction between two pyrimidine bases [47] in single-stranded DNA (ssDNA) and at the flexible ends of poly (dA)-(dT) tracts, but not at their rigid centre [50, 51]. Bending of DNA towards the minor groove reduces CPDs formation [52]. One of the transcription factors, TATA-box binding protein (TBP), promotes the selective formation of 6-4PPs in the TATA-box, where the DNA is bent, but CPDs are formed preferentially at the edge of the TATA box and outside where the DNA is not bent [53]. The amounts of CPDs and 6-4PPs are about 75 and 25%, respectively, of the total UV-mediated DNA damage products [22]; however, the ratio between the yield of CPDs and 6-4PPs mainly depends upon the two adjacent bases involved in the formation of dimers [54]. Thus the heterogeneous distribution of the UV-induced photolesions in the DNA depends on the sequences that facilitate DNA bending as well as the chromatin modulation through the binding of specific protein [55]. Mapping of CPDs in the nucleosome core regions of UV-treated chromatin has revealed the formation of CPDs with an average distance of 10.3 bases, away from the surface of histones [56]. The formation of photoproducts is not restricted to cells exposed to UV-B or UV-B + UV-A radiations; UV-A-induced formation of CPDs has also been observed in bacteria as well as in eukaryotic cells and whole skin [57–59]. Recent studies on the effects of UV-A radiation on rodent and human skin cells have revealed that CPDs are in larger yields than 8-oxo-7,8-dihydroguanine (the most frequent UVA-induced DNA lesion) and DNA strand breaks [48, 60]. The occurrence of 5-methylcytosine-containing photoproducts in UV-irradiated DNA is still controversial. However, Su et al. [61] have reported a new photoproduct of 5-Methylcytosine and Adenine characterized by high-performance liquid chromatography and mass spectrometry. An additional photochemical characteristic for cytosine is the formation of monomeric pyrimidine photoproduct "cytosine photohydrate" (6-hydroxy-5,6-dihydrocytosine) as a result of photohydration reaction (Figure 5) [62]. There is little information concerning the formation of cytosine hydrates in UV-irradiated DNA due to instability of the resulting photoproduct [63]. The oxidation product of pyrimidine bases such as pyrimidine glycols is also formed by means of hydration reaction [42].

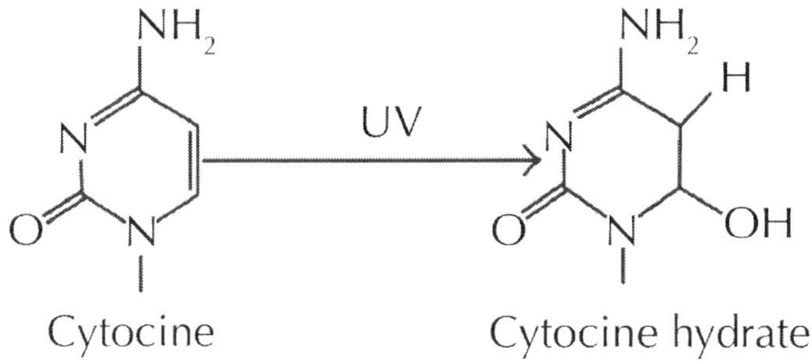

Figure 5: Formation of cytosine photohydrate (6-hydroxy-5,6-dihydrocytosine) as a result of photohydration reaction.

UV-INDUCED PURINE PHOTOPRODUCTS

Although dipyrimidine photoproducts are the preferential outcome of UV-B radiation, the biological importance of UV radiation-induced modifications of DNA purine bases has also been recognized [64]. These comprise the photoproducts that involve, at least, one adenine residue that undergoes photocycloaddition reactions with contiguous adenine or thymine (Figure 6) upon exposure to UV-B radiation [65, 66]. The extent of adenine-containing photoproduct (A-T) is very low (1 × 10−5 in native DNA) but these lesions may contribute to the biological effects of UV radiation in view of the fact that the A-T adduct has been shown to be mutagenic [67, 68]. Photodimerization of adenine (A) involves the cycloaddition of N7-C8 double bond of the 5′-A across the C6 and C5 positions of the 3′-A [65, 69] and generates a very unstable azetidine intermediate. This intermediate photoproduct undergoes competing reaction pathways to form two distinct adenine photoproducts such as adenine dimer (A=A) and Pörschke photoproduct (Figure 6) [70]. Conversion of both these photoproducts into 4,6-diamino-5-guanidinopyrimidine (DGPY) and 8-(5-aminoimidazol-4-yl)adenine (8-AIA), respectively, can be detected from individual acid hydrolysates of UV-irradiated polynucleotides and DNA [71]. It has been found that complexing of UV-irradiated poly(dA)-poly(dT) effectively reduces the formation of A=A photoproduct [71]. Moreover, photoreactivity of adjoining adenine bases in DNA is strongly suppressed by the complementary base pairing [50, 72]. UV-induced ROS acts as a powerful oxidant that may cause oxidative DNA damage. A number of oxidation products of purine bases such as 8-oxo-7,8-dihydroguanyl (8-oxoGua), 8-oxo-Ade, 2,6-diamino-4-hydroxy-5-formamidoguanine (FapyGua), FapyAde, and oxazolone have been reported to form upon exposure of DNA to UV radiation [44, 73, 74].

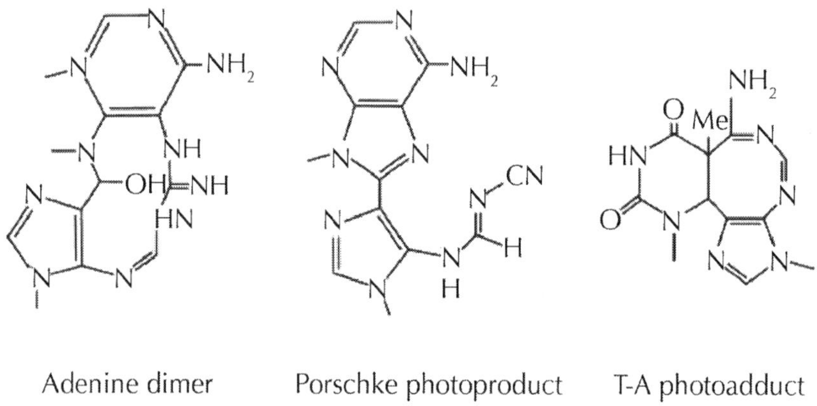

Figure 6: Structure of purinic photoproduct that is, adenine dimer, porschke photoproduct and thymine-adenine photoadduct.

Overall, it has been concluded that UV-induced DNA lesions such as CPDs, 6-4PPs, abasic site, strand breaks, and oxidative product are the predominant and most persistent lesions and if not repaired may cause severe structural distortions in the DNA molecule, thereby affecting the important cellular processes such as DNA replication and transcription, compromising cellular viability and functional integrity and ultimately leading to mutagenesis, tumorigenesis, and cell death [30, 36].

DIFFERENTIAL EFFECTS OF CPDS AND 6-4PPS ON DNA CONFORMATION

UV-induced DNA lesions such as CPDs and 6-4PPs show differential effects on DNA conformation, impairing their regulatory functions and other dynamic processes [24, 26, 75, 76]. Nuclear Magnetic Resonance (NMR) spectroscopy has presented new insights on UV radiation-induced nucleic acid conformation. It has been well established that the comparative orientation of damaged residues is unusual from that observed in unmodified DNA duplexes [75]. Nuclear overhauser enhancement (NOE) study of interactions among the photoadduct H6 and methyl (CH_3) groups has established that the cis-syn CPD changes the cyclobutane conformation from a left-handed twist (observed in the isolated dimer) to a right-handed twist in DNA duplex [46] (Figure 7). Assessment of the chemical shift data suggests that the DNA helix is disturbed more along the 3′- and 5′-side of the cis-syn and trans-syn dimer, respectively. It was revealed that the presence of trans-syn CPD causes distortion to a great extent than the cis-syn product by means of a kink or dislocation at the 5′-side of the dimer in double-stranded DNA [77]. NMR and X-ray diffraction studies of the ultraviolet photoproduct, cis-syn CPD, with the S-cyanoethyl phosphotriester have revealed that the two pyrimidine bases are rotated by −29° base twist, contrasting to the right-handed 36° value observed in B-form DNA [78, 79]. Moreover, in contrast to the cis-syn CPD, the duplex spectra of the trans-syn lesion illustrated no abnormally shifted ^{31}P or imino proton signal, signifying the absence of major distortions in the conformation of the sugar-phosphate backbone [75]. The thymine (T) residues of the CPD form stable wobble pairs with the opposite guanine (G) residues. The T6-G15 wobble pair of the CPD formed hydrogen bonds between the T6-imino and G15-O6 and between the G15-imino and T6-O2. The two T (T5, T6) residues of the CPD in the CPD/GG duplex form wobble base pairs with the opposite G residues, similar to the T6-G15 base pair in the CPD/GA duplex [80]. It has been reported that the preexisting CPDs in the DNA molecule can influence its rotational setting on the histone surface during nucleosome formation [81]. Recently, Rumora et al. [82] have examined the thymine dimer-induced structural changes to the DNA duplex with several small, base-selective reactive chemical probes. The formation of 6-4PPs and their Dewar isomers cause remarkable change in the conformation of DNA duplex. The one- and two-dimensional NMR data on the (6-4)-adduct-containing DNA duplex decamer was analyzed in H_2O and D_2O solutions to elicit the base pairing and unusual conformation in the vicinity of the lesion [76, 83, 84]. The distortion of the double helix caused by the 6-4PP is much greater than that of the CPD [85].

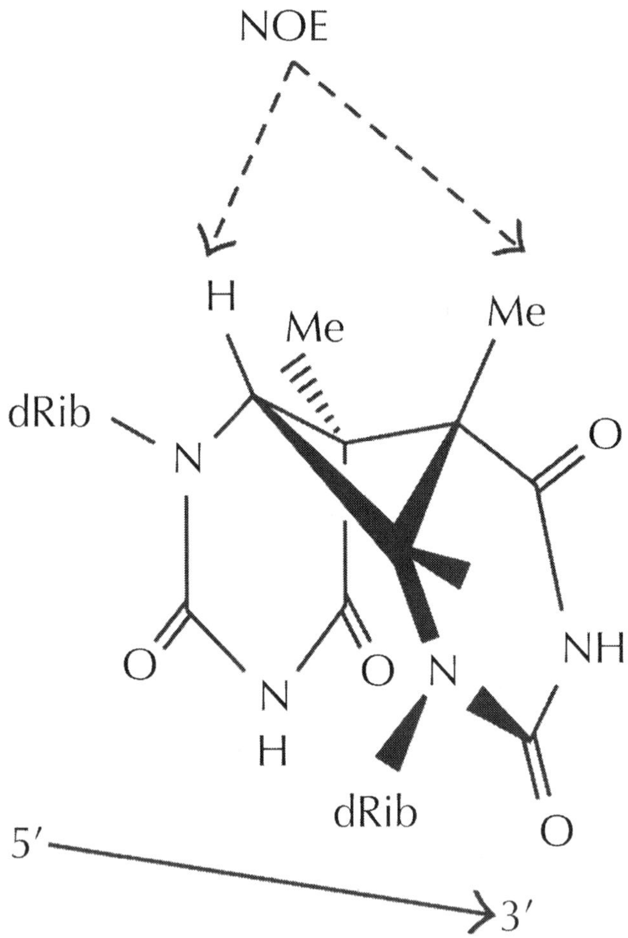

(a)

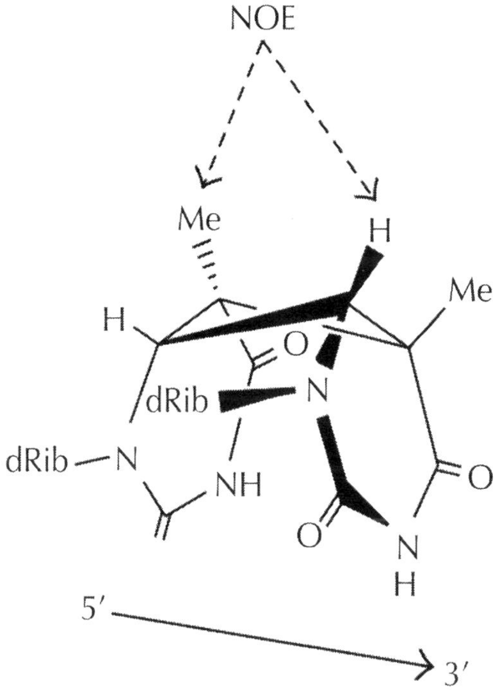

(b)

Figure 7: Cis-syn CPD showing the right-handed or left-handed twist in DNA duplex. Dotted arrows elucidate the strongest nuclear overhauser enhancement (NOE) interaction in both cases (Adopted from Lukin and de los Santos [75]).

The main conformational perturbations caused by the (6-4) adduct and Dewar product are concerned with their effects on global DNA curvature. Both duplex decamers are significantly bent at the lesion sites. In contrast to the 6-4PPs, the 5,6-dihydro-5-hydroxythymine base is the most perturbed part of the 6-4 Dewar lesion. Even though there are no hydrogen bonds between 5,6-dihydro-5-hydroxythymine and its partner adenine residue, this lesion produces minor distortions in comparison to the 6-4PP. In general helical bending induced by the (6-4) adducts and the Dewar product is 44° and 21°, respectively [76]. All the supplementary imino proton resonances from the flanking base pairs were observed in the hydrogen-bonded region, which indicate that the structure of the (6-4) adduct inside the duplex shows a distinctive base orientation due to (6-4) covalent linkage which makes a normal Watson-Crick-type hydrogen bonding unfavorable at the 3′-side of the lesion site with an empty space between the 3′-thymine (T6) and its opposite base (A15) [76]. Contrary to the duplex (6-4) decamer which retains a hydrogen-bonded imino proton at the 5′-side (T5) of the (6-4) lesion, the T5 imino proton of the Dewar lesion is not hydrogen-bonded. The NMR characterization of a 6-4PP dimer containing duplex showed that the 5′-residue of the lesion remains essentially unperturbed. However, the 5′-pyrimidine residue loses aromaticity and acquires an additional hydrophilic group [75, 83]. The glycosyl bond torsion angle at the T5 residue of the (6-4) lesion and the Dewar lesion prevails in the anti and high-anti conformation, respectively, and thus both the lesions exhibit considerable differential effects on DNA backbone conformation. It has been observed that the large structural distortion induced by the (6-4) lesion may ensure a favorable recognition by the repair enzyme, which may possibly elucidate the correlation with the elevated repair rate of the T-T (6-4) adduct than of the T-T Dewar product and the T-T cis-syn dimer [76].

UV-INDUCED DNA DOUBLE STRAND BREAKS

The generation of DNA double strand breaks (DSBs) in UV-irradiated cells, specifically in replicating DNA, has been known for a long time [86]. DNA strand breaks are observed extensively in cells under UV-B irradiation [87, 88]. UV-B-induced ROS [89] as well as DNA lesions (CPDs and 6-4PPs) may cause primary as well as secondary breaks, respectively. These lesions are commonly associated with transcription/replication blockage that may lead to production of DNA double-strand breaks (DSBs) at the sites of collapsed replication forks of CPDs-containing DNA [90, 91] (Figure 8). Dunkern and Kaina [92] also observed UV-C-induced DNA DSBs, arising from replication of damaged DNA. A significantly low amount of DSBs was found in the cell where replication was inhibited. It was assumed that initial photoproducts are converted into DSBs during DNA replication, due to not a distinct process, that is, "collapse of replication forks" [93]. After labeling of replicating DNA of UV-irradiated SV40-transformed human cell lines with radioactive precursors, an increased number of DSBs was observed in NER deficient cells in comparison to NER proficient cells. These results further support the view that DSBs are produced during the replication of unrepaired UV-induced DNA lesions [91]. DSBs can be formed in response to the repair of single strand breaks (SSBs) passing through base excision repair (BER) [94, 95]. Overall, it seems that UV radiation does not directly produce DNA DSBs but rather produces pyrimidine dimers and other photoproducts leading to replication arrest and DSBs. UV-induced replication arrest in the xeroderma pigmentosum variant (XPV) followed by the accumulation of Mre11/Rad50/Nbs1 complex and phosphorylated histone H2AX (γ-H2AX) in large nuclear foci at sites of stalled replication forks also suggests that UV damage leads to the formation of DSBs during the course of replication arrest [90, 95].

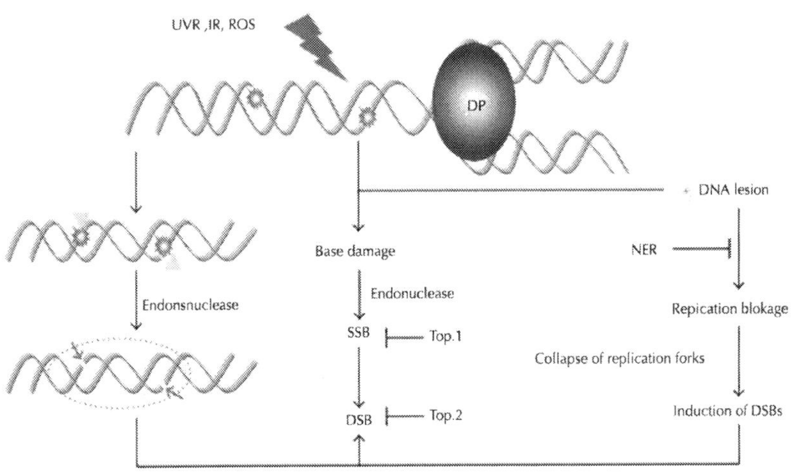

Figure 8: Schematic representation showing different pathways of DSBs.

A number of pathways have been considered for the formation of DSBs at a stalled replication fork. It was shown that when the DNA replication machinery encounters a replication-blocking lesion, DNA polymerase (DP) enzyme is stalled at the blocked site resulting in the formation of a Y-shaped DNA structure, which may be recognized by a specific endonuclease, that successively makes a nick in the template strand resulting in the induction of a DSB close to the replication-blocking lesion [144]. Furthermore, replication stresses may trap topoisomerase I (Top1) cleavage complexes leading to generation of DSBs by preventing Top1-mediated DNA religation [145]. Free radicals may also cause DSBs [146] by preventing the topoisomerase II (Top2)-mediated DNA religation [144, 147]. Recently, Harper et al. [148] have shown that radiation-induced SSBs and non-DSB DNA damage contribute to the formation of replication-induced DSBs. In spite of the above possible facts regarding the formation of DSBs, more experimental evidences are still needed.

DETECTION OF DNA DAMAGE

Several workers have attempted to detect different types of DNA lesions and presently a number of detection strategies are widely used (Table 1). An alkaline gel method for quantitating single-strand breaks (SSBs) in nanogram quantities of nonradioactive DNA was developed by Freeman et al. [149]. Mitchell et al. [150] have developed a method for the detection of CPDs, where DNA is labeled with radioactive substances followed by agarose gel electrophoresis and densitometric analysis and finally digesting with endo. III and endo.V before analyzing on sequencing gels. UV-B induced DNA damage in mammalian genome was reported by Wang et al. [96] using the PCR-based short interspersed DNA element- (SINE-) mediated detection method. For analyzing the 6-4PPs, terminal transferase-dependent PCR (TD-PCR) has been used [97]. UV-induced decrease in template activity of genomic DNA of cyanobacterium Anabaena strain BT2 was documented by Kumar et al. [98] using the PCR-based assays such as random amplified polymorphic DNA (RAPD) and rDNA amplification. Similarly, UV-B-induced DNA damage was also detected in Anabaena variabilis PCC 7937 and Rivularia sp. HKAR-4 by PCR (data not published). The formation of thymine dimer (T^T) within human genomic DNA has been detected by immunocoupled PCR (IC-PCR) [99]. DNA damage such as SSBs, DSBs, and oxidative DNA damage caused by UVR, ultrasound electromagnetic frequency radiation, and so forth may be detected by comet assay [100]. Recently, a modified version of comet assay (apo/necro-comet assay) has been developed that differentiates viable, apoptotic, and necrotic cells and also correlates the DNA fragmentation pattern [101]. Both single and DSBs as well as apoptosis can also be detected by TUNEL assay [104, 105]. However, it has been experienced that TUNEL assay is not able to distinguish various types of cell death; hence, an alternate method based on flow cytometry (FCM) has been developed for the detection of apoptosis [109]. Recently, apoptosis in tumor cells caused by X-rays has been analyzed using $^{125-}$I-labeled annexin V [111]. FCM assay is useful in detecting chromosomal aberrations, sister chromatid exchange, chemical adducts to DNA, and DNA strand breakage [110]. Alkaline unwinding FCM (AU-FCM) may be used to detect nucleotide excision repair (NER) [151]. The changes in DNA organization in the individual cells can be determined by halo assay [102]. SSBs at the single cell level can be assessed by alkaline-halo assay (AHA), where cells are embedded in melted agarose and spread on the microscope slide and then incubated in a high-salt alkaline lysis solution followed by another incubation in a hypotonic alkaline solution and, finally, stained with ethidium bromide (EtBr). Under these conditions, single-stranded DNA fragments diffuse radically from the nuclear cage [103]. DNA strand breaks (SSBs, DSBs, and alkali-labile sites) induced by genotoxic agent such as UVR can also be detected by fluorometric analysis of DNA unwinding (FADU) assay, which was first reported by Birnboim and Jevcak [116] to detect X-ray-induced DNA damage in mammalian cells. Numerical aberrations in chromosome can be detected efficiently by fluorescence in situ hybridization (FISH) method [107]. Recently, immuno-dot-blot assay is used extensively to detect UV-induced photoproducts in various organisms such as mammals, cyanobacteria, phytoplankton, macroalgae, and liverwort [17, 59, 112, 113]. This technique is based on use of thymine-dimer specific antibodies followed by blotting and chemiluminescence method. Another detection strategy includes radio-immunoassay (RIA) which is used to estimate CPDs and 6-4PPs [114]. The very low amount of CPDs caused by UVR in bacterioplankton and marine viruses may be detected very efficiently using RIA method [115]. Kara et al. [152] have studied the electrochemical detection of DNA damage by direct and indirect irradiation with radioactive technetium (TC-99m) and iodine (I-131). Certain photoproducts such as 5-Methylcytosine and adenine can be detected by high-performance liquid chromatography and mass spectrometry [61]. Recently, Kumari et al. [108] have made an attempt to dissect various strategies for detection of DNA lesions produced by a number of genotoxic agents.

Table 1: Various strategies for the detection of damaged DNA

DNA damage detection strategies	Types of lesions detected	References
PCR based assay (TDPCR, LMPCR, ICPCR, SINE)	Decrease in DNA template activity, T < > T CPDs, 6-4PPs	[96–99]
Commet assay (Single-cell gel Electrophoresis)	Oxidative DNA damage and single/double strand break	[100, 101]
Halo assay/AHA/ FHA	Chromatin fragility/single strand breaks at the single cell level	[102, 103]
TUNEL assay	Single/double strand breaks, apoptosis	[104, 105]
HPLC-MS/MS	Oxidative DNA damage; CPDs, 6-4PPs and their related Dewar valence isomers; 5-hydroxy-2-deoxyuridine,8-oxo-7,8-dihydro-2-Deoxyadenosine; 5-Methylcytosine and adenine	[48, 54, 61,106]
FISH	Chromosomes with numerical aberrations	[107, 108]
FCM	Chromosomal aberrations, sister chromatid exchange, chemical adducts to DNA and DNA strand breakage	[109, 110]
Annexin V labeling	Chromatin condensation, DNA fragmentation, radiation-induced apoptosis	[109, 111]
Immuno-dot-blot assay	CPDs, 6-4PPs and their Dewar valence isomers	[17, 59,112, 113]
RIA	CPDs and 6-4PPs	[114, 115]

GC-MS	Strand break, modified bases, abasic sites, DNA-protein crosslinks and other oxidative DNA damage.	[39]
FADU	Single/double strand breaks and alkali-labile sites	[88, 116]
NMR spectroscopy	Lesions induced distortions of DNA duplex	[78–80]

PCR: polymerase chain reaction; TDPCR: terminal transferasedependent PCR; LMPCR: ligation-mediated PCR; ICPCR: immuno-coupled PCR; SINE: short interspersed DNA element; AHA: alkaline-halo assay; FHA: fast halo assay; TUNEL: terminal deoxyribonucleotidyl transferase mediated deoxyuridine triphosphate nick end labeling; HPLC-MS/MS: high-performance liquid chromatography coupled to tandem mass spectrometry; FISH: fluorescence in situ hybridization; FCM: flow cytometry; RIA: radio immunoassay; ELISA: enzyme-linked immunosorbent assay; GC-MS: gas chromatography-mass spectrometry; NMR: nuclear magnetic resonance spectroscopy.

DNA REPAIR

The idea about the ability of living beings to overcome the lethal effects of UV-radiation emerged as early as the mid 1930s [153], but the existence of repair mechanisms was observed by Kelner [154] and Dulbecco [155] independently. The determination of a particular repair pathway within the cell mainly depends on the types and location of lesions in the genome [156]. The biochemical and molecular studies on repair pathways have been extensively investigated in some model organisms such as E. coli, S. cerevisiae, and human, where specialized repair proteins scan the genome continuously and encounter the DNA lesions by triggering several distinct repair mechanisms such as photoreactivation, excision repair (BER and NER), mismatch repair (MMR), and some specialized forms of repair system such as SOS response, damage tolerance, and apoptosis.

PHOTOREACTIVATION

The process of photoreactivation is executed by means of a photoreactivating enzyme known as "photolyase", which is well conserved and found throughout the three domains of life (Table 2). The enzyme binds specifically to the CPDs (CPD photolyase) or 6-4PPs (6-4 photolyase) and directly monomerizes the cyclobutane ring of the pyr <> pyr, using the energy of visible/blue-light and protects the genome from deleterious effects of UVR [157, 158]. The absorption of every blue-light photon may split approximately one dimer [159]. CPD photolyases have been reported in diverse groups such as archaea, bacteria, fungi, virus, plants, invertebrates, and many vertebrates including aplacental mammals (Table 2). On the other hand, 6-4 photolyases have been identified in certain organisms like Drosophila, silkworm, Xenopus laevis, and rattle snakes [22]. Photolyases seem to be absent or nonfunctional in placental mammals like human [118, 132, 160]. However, Sutherland [161], Sutherland and Bennett [162], and Harm [163] have demonstrated photolyase activity in cells and tissues, including white blood cells (WBCs) of several placental mammals, such as humans, ox, cat, and mouse. A number of workers have identified a human photolyase which shows homology with Crygene (plant blue-light receptor) and about 40% sequence identity to the Drosophila 6-4 photolyase, but their exact roles in repair process, whether it acts as a photolyase or as a photoreceptors, are still under investigation [164].

Table 2: Photolyase enzymes in four different kingdoms

Kingdom	Organism	CPD Photolyase	6-4PP Photolyase	References
	Halobacterium halobium	+	–	[117]
Archaebacteria	Methanobacterium thermoautotrophicum	+	–	[118, 119]
	Sulfolobus tokodaii	+	–	[120]
	Bacillus firmus	+	–	[121]
	Escherichia coli	+	–	[122, 123]
	Salmonella typhimurium	+	–	[122]
Eubacteria	Anacystis nidulans	+	–	[124]
	Synechocystis sp. PCC 6803	+	–	[125]
	Streptomyces griseus	+	–	[126]
	Myxococcus xanthus	+	–	[127]

	Vibrio cholerae	+	–	[128]
	Saccharomyses cerevisiae	+	–	[122]
	Neurospora crassa	+	–	[129, 130]
	Drosophila melanogaster	+	+	[131]
	Homo sapiens	?	?	[118, 132]
	Carassius auratus	+	–	[133]
	Oryzias latipes	+	–	[118]
Eukaryotes	Monodelphis domestica	+	–	[134]
	Potorous tridactylis	+	–	[118]
	Xenopus laevis	+	+	[135]
	Arabidopsis thaliana	+	+	[136]
	Chlamydomonas reinhardtii	+	–	[137]
	Cucumis sativus	+	–	[138, 139]
	Ginkgo biloba	+	–	[140]
	Medicago sativa	+	–	[20]
	Triticum sp.	+	–	[141]
	Fowl pox virus (FPV)	+	–	[142]
Viruses	Melanoplus sanguinipes entomopox virus (MsEPV)	+	–	[143]
	Chrysodeixis chalcites nucleopolyhedrovirus (ChchNPV)	+	–	[106]

DNA photolyases (45–66 kDa) having 420–616 amino acid residues [158] are monomeric flavin-dependent repair enzymes, consisting of two known cofactors, a catalytic cofactor and a light-harvesting cofactor. Till date, 5,10-methenyltetrahydrofolate (MTHF) [122], 8-hydroxy-5-deaza-riboflavin (8-HDF) [124], and FMN [165] are known as light-harvesting cofactors, which absorb light energy efficiently and transfer them to $FADH^-$ [166]. Deprotonated reduced flavin adenine dinucleotide ($FADH^-$) is found in all known photolyases as a catalytic cofactor, which transfers energy in the form of an electron to the CPD, splitting the cyclobutane ring with the generation of two monomeric bases [157, 167] (Figure 9).

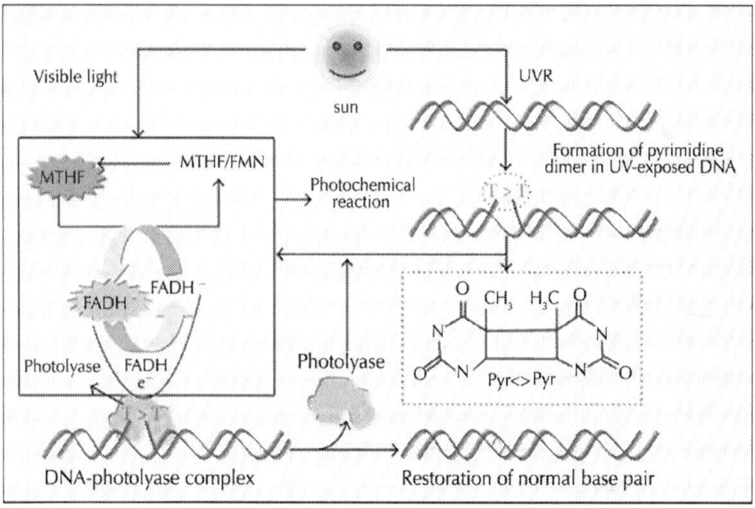

Figure 9: Photoreactivation: incidence of ultraviolet radiation (UVR) results in pyrimidine lesion (thymine dimer), which is recognized by a photoreactivating enzyme "photolyase". The light energy (>380 nm) is trapped by the antenna molecules of photolyase (such as MTHF/8-HDF/FMN) and transfers them to catalytic cofactor $FADH^-$which becomes excited and transfers energy to the pyrimidine dimer in the form of e^-, splitting the CPD into two monomeric unit, and then electron is transferred back to the flavin molecule.

In comparison to other eukaryotic systems, reports on the repair of UV-induced DNA damage in plants are still very limited. To avoid the deleterious effects of UVR, plants have acquired two main protective strategies; shielding by flavonoids and phenolic compounds [168, 169] and DNA repair by photoreactivation. Photoreactivation mediated by the enzyme photolyases is thought to be the major DNA repair pathway in several higher plants such as rice, Arabidopsis, wheat, and maize [170–172]. Studies on Arabidopsis seedling, rice, and alfalfa indicate that photoreactivation greatly enhances the rate of removal of dimers, although, in the absence of photoreactivating (blue) light, dimers are slowly eliminated from bulk DNA and 6-4PPs are generally observed to be repaired more quickly than CPDs [173, 174]. Plants grown in the presence of photoreactivating radiation can eliminate the majority of both 6-4 products and CPD lesion within hours, or in some cases minutes, of their induction [30]. The structural information about the interaction between CPD lesions and photolyases became clear with the help of X-ray crystallography [175] and nuclear magnetic resonance (NMR) spectroscopy [176]. However, how DNA photolyases find lesions in the DNA molecule is still not clear [158]. It has been observed that about 240 KJ/mol of energy is captured upon absorption, out of which about 125 KJ/mol energy is consumed during the initial electron transfer from the excited FADH to CPD lesions [177]. The splitting of CPD lesion proceeds rapidly within 0.6 nanosecond [167, 178]. The back-transfer of electrons from the CPD lesion to the FADH radical is efficiently avoided by the enzyme before completion of cleavage of the cyclobutane ring [157]. With the help of ultrafast femtosecond laser spectroscopy, MacFarlane and Stanley [178] have suggested that photolyase enzyme is indeed left in the semiquinonid state after accomplishment of repair of the CPD lesion. However, Kavakli and Sancar [179] have analyzed the role of intraprotein electron transfer in photoreactivation by DNA photolyase and found that photoreduction process is not a regular part of the photolyase photocycle under physiological conditions, because the enzyme may undergo at least 25 repair cycles before loosing its activity. After completion of DNA repair, a thymine pair is flipped back into the duplex DNA to form a hydrogen bond with their complementary adenine base. Fourier transform infrared spectroscopy (FTIR) has revealed that the relaxation of DNA backbone proceeds very slowly than the repair of CPD lesions [180]. In the absence of photoreactivating light, the enzyme binds to and stimulates the removal of UV damage by stimulating the NER system in vivo or in vitro and defense against DNA damage even in the absence of light [181].

EXCISION REPAIR

Unlike photoreactivation, excision repair is a multistep, dark repair pathway, where an abnormal or damaged base is removed by two major subpathways: (i) base excision repair (BER) and (ii) nucleotide excision repair (NER).

Base Excision Repair (BER)

BER is the predominant DNA repair pathway against base lesions arising from hydrolytic deamination, strong alkylating agents, ionizing radiation (IR), or by different intracellular metabolites and, indirectly, also by UV radiation via generation of ROS [182–184] and proceeds through a series of repair complexes that act at the site of DNA damage [185, 186]. The efficiency and specificity of the repair pathway are determined by several forms of DNA glycosylase which removes different types of modified bases (Table 3) by cleaving the N-glycosidic bond between the abnormal base and deoxyribose creating either an abasic site or an SSB [187]. Recently, Parsons et al. [188] have discovered that the formation of DNA repair complexes on damaged DNA stabilizes BER proteins. On the contrary, BER proteins that are not involved in repair are ubiquitylated by carboxyl terminus of Hsc70 interacting protein (CHIP) and subsequently degraded by the proteasome.

Table 3: DNA glycosylases and their probable substrate in bacteria, yeast, and human (modified from Sinha and Häder [22])

Glycosylases	Organisms	Genes	Substrate
(1) Uracil DNA glycosylase	E. coli	Ung	uracil from ss- and ds-DNA
		Dug/mug	U from U:G, ethenocytosine, hypoxanthine and 5-hydroxycytosine
		Dut	?
	S. cerevisiae	UNG	Uracil
	Human	UNG1/ UNG2	uracil from ss- and ds-DNA
		SMUG1	uracil from ss-DNA, hydroxymethyluracil, formyluracil
(2) 3-methyl adenine DNA glycosylase	E. coli	tag	3-methyladenine
		alkA	3-methyladenine, 7-methylguanine, 2-methylcytosine, 5-formyluracil
	S. cerevisiae	MAG1	3-methyladenine, 7-methylguanine
	Human	MPG Aag	3-methyladenine hypoxanthine

(3) UV-endonuclease	T4	den V	cis-syn-cyclobutane-type pyrimidine dimer
	E. coli	?	?
	Bacillus subtilis	UVDE	Pyrimidine dimers
	S. cerevisiae	?	?
	S. pombe	UVDE	Pyrimidine dimers
	Human	?	
(4) Endonuclease III/thymine glycol DNA glycosylase	E. coli	Nth	5-hydroxycytosine, thymine glycol, urea
	S. cerevisiae	NTG1	oxidative DNA damage, thymine glycol and formamido-pyrimidines, oxidized pyrimidines, 2 formamido-pyrimidine-G, Me7- formamido-pyrimidine-G
		NTG2	oxidative DNA damage, Thymine glycol and formamido-pyrimidines residues, 5-hydroxycytosine, oxidized pyrimidines, Me7-fapy-G
	Human	NTH1	oxidized guanine lesions
(5) Endonuclease VIII	E. coli	nei	Thymine, thymine glycol, urea, 5-hydroxycytosine, dihydrothymine, and b-ureidoisobutyric acid
	S. cerevisiae	?	?
	Human	NEIL1	5-hydroxyuracil, 5-hydroxycytosine, 5,6-dihydrouracil, thymine glycol, formamido-pyrimidines (FapyA/G)
		NEIL2	5-hydroxyuracil and 5-hydroxycytosine
		NEIL3	?

(6) fapy/8-oxoguanine DNA glycosylase	E. coli	fpg/mutM	2,6-diamino-5-formamidopyrimidine 8-oxo-7,8-dihydroguanine, 5-hydroxycytosine
	S. cerevisiae	OGG1	2,6-diamino-5-formamidopyrimidine 8-oxoG, 2 formamidopyrimidine-G, Me7-formamidopyrimidine-G
	Human	hOGG1	8-hydroxyguanine, Me7-formamidopyrimidine-G
(7) A-G-mismatch DNA glycolsylase	E. coli	mut Y	Adenine/C
	S. pombe	spMYH	2-aminopurine/G and A/2-aminopurine, Adenine/C
	Human	MYH	Adenine from G:A, 8-oxoG:A, 2-hydroxyadenine
(8) G-T-mismatch DNA glycosylase	E. coli	?	?
	M. thermoautrophicum	Mig-Mth	Thymine residues from T-G mismatches
	S. pombe	thp1	Uracil from G:U
	Human	MBD4 (≈MED1)	Thymine from T:G
		TDG	Recognizes a G:T mispair in a CpG sequence
(9) Formyluracil DNA glycosylase	E. coli	mug	Formyluracil mispaired with A & G
		mutM	?
		mutS	Formyluracil mispaired with G
	S. cerevisiae	?	?
	Human	MBD4	Formyluracil mispaired with G
		hNTH1	?
(10) Hydroxymethyl uracil DNA glycosylase	E. coli	?	?
	S. cerevisiae	?	?
	Human	?	5-hydroxymethyluracil mispaired with G

The extent of BER conservation among E. coli and mammals has led to progress in our understanding of mammalian BER and here, a general overview of the mammalian BER pathway will be discussed.

As a result of multiple interactions with a number of repair proteins XRCC1 plays a crucial role in the coordination of BER and SSB repair [189]. The interaction between XRCC1 and polymerase β (Pol. β) and its functional aspects was confirmed after UV-A-induced oxidative damage in living mammalian cells [190]. The AP-site is removed by the action of AP endonuclease-1 (APE-1) along with phosphodiesterase that breaks the DNA strand along 5′ or 3′ to the AP site, respectively, and subsequently the gap is filled by a repair DNA polymerase and the strand is joined by a DNA ligase (Figure 10) [182, 191]. It has been reported that the repair Pol. β itself has the capacity to excise the 5′ deoxyribose phosphate residues, that is, generated by the combined actions of DNA glycosylase and ClassII AP endonuclease [192]. The major APE-1 that was discovered independently as an abasic site-specific endonuclease homologous to the E. coli Xth protein [193] incises duplex oligonucleotides containing 5,6-dihydroxyuracil (DHU), 5-hydroxyuracil (5-ohU), and alpha-anomeric 2′-deoxynucleosides (i.e., αdA andαT) residues in human cells [194].

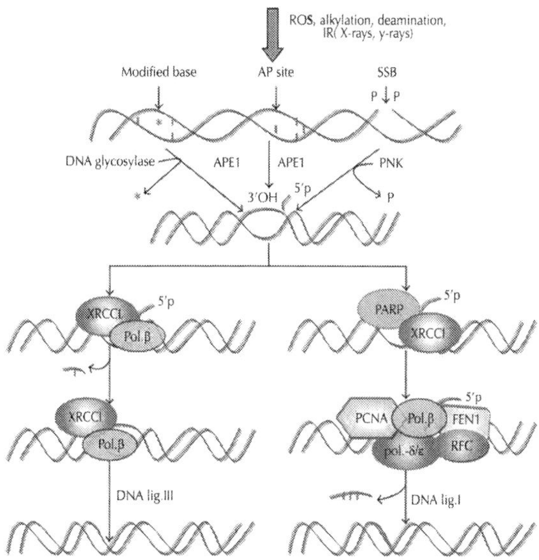

Figure 10: Schematic overview of mammalian SP-BER (a), and LP-BER (b). SP-BER is initiated by the activity of glycosylase and APE1, followed by scaffold protein XRCC1 and pol. β to remove the damaged

nucleotide and DNA ligase III seals the nick. In case of LP-BER, after DNA damage by ionizing radiation, PNK is recruited to convert the damaged ends to 3′OH and 5′P moieties. Here PARP1/2, followed by XRCC1, is involved. PCNA and DNA pol. β and/or pol. −δ/ε extend and fill the gap by >2 nucleotides. Replication factor-C (RFC) is required to load the PCNA on DNA. Ultimately the resulting 5′flap of DNA is removed by the flap endonuclease I (FEN1) and subsequently the nick is sealed by DNA ligase I.

DNA having one nucleotide lesion is removed by short-patch BER (SP-BER) whereas two/more nucleotide lesion is repaired by long-patch BER (LP-BER) pathway [195] (Figure 10). Recently, Almeida and Sobol [184] have proposed a unified model of SP-BER and LP-BER. On the basis of measuring the BER efficiency and presence of a single modified base in a plasmid molecule transfected into mammalian cells, Sattler et al. [196] made the first attempt to verify the occurrence of LP-BER in vivo. It is assumed that majority of repair takes place through SP-BER, initiated either by monofunctional or by bifunctional glycosylase [184]. The pathway of SP-BER after excision of damaged base involves the recruitment of poly (ADP-ribose) polymerase-1 (PARP-1) followed by scaffold protein XRCC1 and DNA pol. β to replace the damaged nucleotide. DNA ligase III (Lig. III) seals the nick and restores the intact DNA.

It has been observed that radiation- (X-rays, γ-rays) induced breaks exist mainly as 5′p and 3′p at the margin of the gap [197, 198] which is converted by the polynucleotide kinase (PNK) or APE1 into 5′P and 3′OH moieties, necessary for the DNA synthesis [199]. Unlike SP-BER, LP-BER involves proliferating cell nuclear antigen (PCNA) coupled with DNA pol. −δ/ε or β which extends and fills the gap by inserting 2–13 nucleotide [184]. The replication factor C (RF-C) is required to load PCNA onto the damaged DNA [195]. The flap endonuclease (Fen1) protein then displaces the ensuing DNA flap leaving a nick which is ligated by DNA ligase I (Lig. I) [200, 201].

In several plant species, some of the genes requisite for dark repair have been identified [171, 172, 202–204]. The available evidence supports the additional existence of enzyme-mediated excision-repair mechanisms in a variety of systems including pollen, whole seedlings and plants, and protoplasts derived from leaves and cultured cells [205, 206]. The formation of AP sites has been observed in seeds of

Zea mays during early germination. This phenomenon was attributed to the action of DNA glycosylases on lesions accumulated during seed storage [207], implying the presence of BER in this species. Likewise, an enzyme activity attributed to uracil-DNA glycosylase was found in cultured cells of Daucus carota [208]. Recently, it has been reported that the mechanisms of BER and NER (but not photoreactivation) in higher plants are active in proliferating cells [172].

Nucleotide Excision Repair (NER)

NER is critically important in the repair of UV-induced DNA lesions and is one of the most versatile and flexible repair systems found in most organisms but highly conserved in eukaryotes. It sorts out a wide range of structurally unrelated DNA lesions, such as CPDs and 6-4PPs caused by UVR, bulky chemical adducts, DNA-intrastrand crosslinks, and some forms of oxidative damage, that cause helical distortion of the DNA double helix as well as modification of the DNA chemistry and interfere with DNA duplication and transcription [209,210]. Although both 6-4PPs and CPDs are removed presumably by the same NER proteins, the relative repair efficiency of both of these lesions varies considerably in mammalian cells. It has been established that in human and hamster cells, the elimination of 6-4PP is at least fivefold faster than that of CPDs [211].

Discovery of NER was first described in E. coli [212, 213] where about six proteins such as UvrA, B, and C (known as ABC-complex, which shows excinuclease activity), UvrD (helicase II), DNA polymerase I (pol. I), and DNA ligase are recruited to complete the repair [214, 215]. Eukaryotic NER is known to be similar to prokaryotes regarding the biochemical strategy used but differs widely in the nature and number of proteins used [210]. The eukaryotic NER pathway has extensively been studied at the molecular level in yeast and human cells. A schematic representation of the NER pathway in human is illustrated in Figure 11.

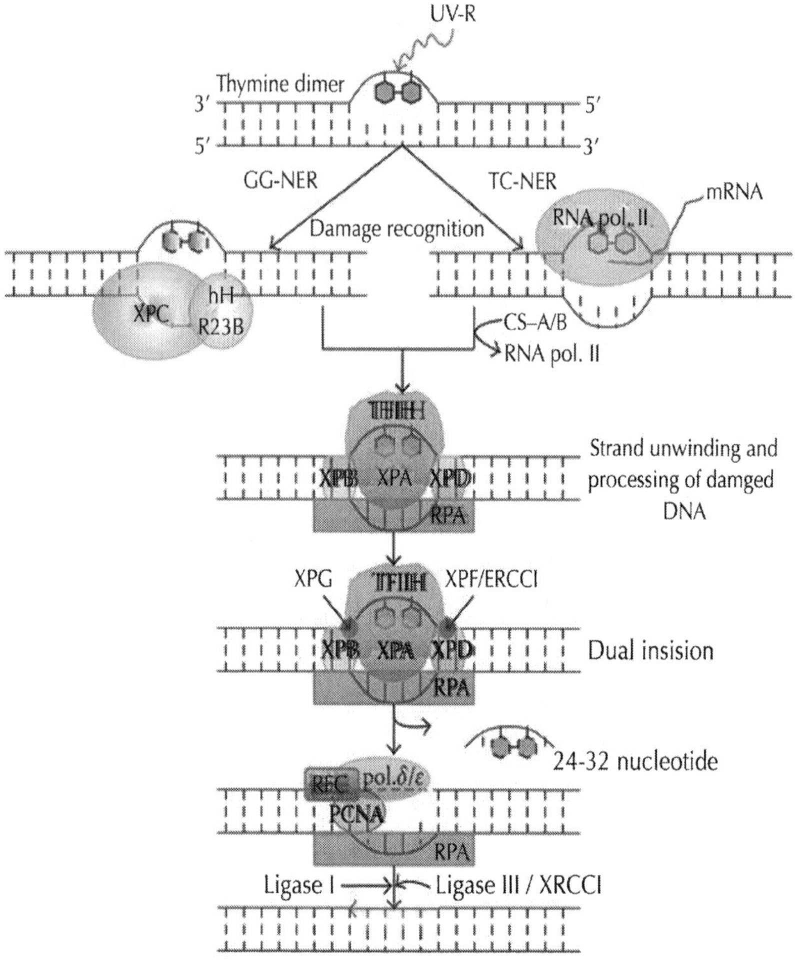

Figure 11: Molecular mechanisms of global genome nucleotide excision repair (GG-NER) and transcriptional coupled nucleotide excision repair (TC-NER) in mammals. For details see the text.

NER can be subdivided into differentially regulated subpathways such as global genome NER (GG-NER) and transcription-coupled NER (TC-NER): repair of lesions over the entire genome, referred to as global genome repair (GGR), and repair of transcription-blocking lesions present in transcribed DNA strands, referred to as transcription coupled repair (TCR). Both repair systems removed a wide range of UV-induced

DNA lesions in a sequential way that includes damage recognition, opening of DNA double helix at damage site, and dual incisions on both sides of the lesion followed by resynthesis and ligation [210, 216] (Figure 11). In human XP-C cells, where the removal of 6-4PP takes place through the TCR pathway, the repair seems to be threefold slower as compared to normal cells. This indicates that the GGR is the most preferred and efficient pathway for 6-4PP removal [211].

MECHANISTIC DIFFERENCES BETWEEN GG- AND TC-NER IN MAMMALIAN CELLS

Global genome repair (GGR) is a random process that occurs gradually, whereas TCR, which is firmly linked to RNA polymerase II (RNA pol II) transcription, is highly specific and efficient. NER defects are associated with a surprisingly wide clinical heterogeneity.

It is assumed that TC-NER proceeds when the transcription machinery RNA pol II encounters a lesion. To progress the transcription-coupled repair (TCR), the stalled polymerase must be displaced, which is brought about by the recruitment of two proteins CSA and CSB. The CSA protein (44 kDa) which belongs to "WD repeat" family of proteins exhibits structural and regulatory roles and CSB proteins (168 kDa) which belong to SWI/SNF family of proteins exhibit DNA-stimulated ATPase activity [217–219]. As stated earlier, elongation of active RNA pol II is prerequisite for efficient TCR; the CSA and CSB gene products are required for efficient repair only during the elongation stages of RNA pol II transcription. It has been suggested that the RNA pol II backs up some nucleotides upon encountering the lesion to facilitate the accessibility of the repair machinery to the lesion site [220, 221]. It is expected that the CSB protein ubiquitinates the stalled elongating RNA pol II complex at the lesion and enhances the assembly of repair factors [222]. However, the fate and the role of ubiquitylated RNA pol II have yet to be clarified [223]. Recently, Fousteri et al. [224] has revealed that CSB is a prerequisite factor in vivo to assemble NER proteins while it is not essential to recruit TFIIH or NER complex in vitro. In living human cells, Proietti-De-Santis et al. [225] have shown that CSB is required during the first phases of RNA pol II transcription initiation. At higher dose of UV radiation (i.e., used for 6-4PP detection), elongation of RNA pol II is greatly impaired, affecting the efficiency of TCR. Hence, at higher UV radiation the GGR overrules the TCR pathway [55]. The probable relationship between TCR and blockage of RNA transcription subsequent to UV-irradiation may possibly be due to CSB which acts as a chromatin remodeling factor at both levels, allowing either the recruitment of the transcription machinery at the initiation sites or the remodeling of the stalled RNA polymerase allowing the NER factors to access the lesion [223].

In GG-NER pathway, lesions produced in transcriptionally silent areas of the genome are recognized by hHR23B-XPC protein complex in an energy-independent manner. The rate of GGR strongly depends on the type of lesion. For instance, 6-4PPs are removed much faster from the genome than CPDs, possibly because of disparity in affinity of the damage sensor hHR23B-XPC. XPC is the sole XP factor not essential for TCR and is restricted to GGR [211]. It is supposed that XPC binds preferentially to the stretch of ssDNA that occurs in the nondamaged strand, opposite to a lesion [252]. However, association of UV-damaged DNA-binding protein (UV-DDB) with a cullin-based ubiquitin ligase has revealed novel mechanistic and regulatory aspects of mammalian GG-NER. It was reported that XPC and UV-DDB materialize to assist for the efficient recognition of UV-induced photolesions and that both factors are ubiquitylated [250, 253]. Lesions that cause little distortion can be recognized by the DDB complex which is also part of an E3 ubiquitin (Ub) ligase that poly-ubiquitinates XPC and XPE [216]. A mechanistic pathway for recruitment of XPC complex to the major UV-induced photolesions (i.e., CPDs and 6-4PPs) has recently been elucidated [216, 253]. It was shown that DDB complex is recruited first to the lesion (CPD) before the XPC complex, on little distorted DNA helix; however, in case of large distortion of the DNA helix caused by 6-4PPs, direct recognition by XPC is also possible for this lesion (Figure 12) [216, 253]. However, the method by which XPC locates a lesion in the vast excess of undamaged DNA in the enormous mammalian genome is not clear [254] and it needs more investigation. On the basis of DNase I footprinting, Sugasawa et al. [226] showed that hHR23B-XPC attaches directly to DNA damage and alters the DNA conformation around the lesion. The XPC protein (125 kDa) is complexed with hHR23B protein (58 kDa). These two proteins are human homologs of the yeast (S. cerevisiae) NER factor Rad4 and Rad23, respectively

(Table 4). In mammalian cells, the quantity of hHR23B is higher than the XPC [255] and in vitro activation of the later protein is stimulated by hHR23B possibly in a structural quite than a catalytic way [256, 257]. It has been assumed that hHR23A can substitute for hHR23B in complex formation and stimulation of XPC repair activity [258]. Both hHR23A and -B harbor a ubiquitin-like moiety at their amino terminus [230]. hHR23B-XPC complex or only XPC exhibits a similar high affinity for both UV-induced single-stranded DNA (ssDNA) and double-stranded DNA (dsDNA) [230, 259], preferentially binds to DNA with various lesions [259] and even to small bubble structures with or without a lesion [260]. hHR23B-XPC is absolutely required for dual incision as well as for open complex formation during GG-NER [243, 261]. Overall, it has been distinguished that RNA pol II with the recruitment of two proteins (i.e., CSA and CSB) proceeds the TCR while hHR23B-XPC complex is the first factor in NER that initiates GGR by sensing and binding lesions, locally distorting the DNA double helix and recruiting the other factors of the system [227].

Table 4: NER proteins and their probable role in human and S. cerevisiae

NER Factors	Name			Function	References
	Human	Yeast S. cerevisiae	Size (a.a)		
XPC-hHR23B	XPC	Rad4	940	Binds damaged DNA; recruits other NER proteins; works with hHR23B. involved only in GGR	[22, 226, 227]
	hHR23B	Rad23	409	Stimulates XPC activity in vitro; contains ubiquitin domain	[22, 227–229]
	hHR23A	Rad23	363	Can substitute for hHR23B in vitro	[227, 230]
	CEN2	—	172	Stabilizes the XPC-hHR23B complex	[216, 231]
	Mat1	TFB3	309	CDK assembly factor; CAK subcomplex	[227, 228]
	Cdk7	Kin28	346	CDK, C-terminal domain kinase; (CAK) subcomplex; phosphorylates RNA pol. II and other substrates	[227, 228]
	Cyclin H	CCL1	323	Cyclin; CAK subcomplex	[227, 228]
	TFB5/ TTDA (p8)	Tfb5	71	Stabilizing subunit	[228, 237]
XPA	XPA	Rad14	273	Binds damaged DNA and facilitates repair complex assembly; affinity for ssDNA	[183, 238,239]
	RPA70	Rfa1	616	Stabilizes opened DNA complex; positions nucleases; ssDNA binding	[227,240–242]

RPA	RPA32	Rfa2	270	Stabilizes opened DNA complex; positions nucleases; ssDNA binding	[227,240–242]
	RPA14	Rfa3	121	Stabilizes open complex (with XPA/Rad14)	[22, 240, 241]
XPG	XPG	Rad2	1186	Endonuclease (catalyzes 3 incision); stabilizes full open complex	[228,243–245]
ERCC1-XPF	ERCC1	Rad10	297	Part of structure-specific endonuclease; catalyzes 5 incision; interstrand cross-link repair	[227,246–248]
	XPF	Rad1	905	Part of endonuclease (5 -incision); recombination via single-strand annealing	[227, 246,249]
DDB	DDB1	—	1140	CPD recognition?	[228, 250,251]
	DDB2	—	428	Chromatin remodeling?	

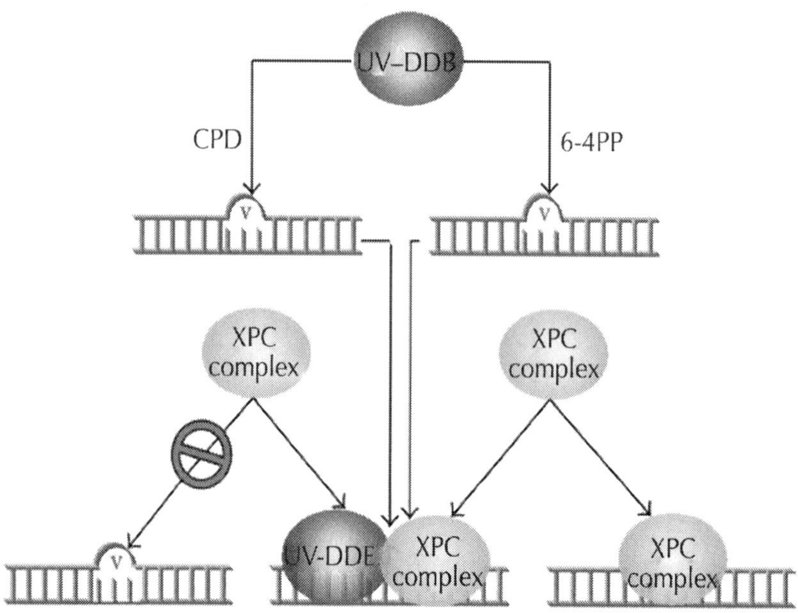

Figure 12: Different pathway for recognition of DNA lesions such as CPD and 6-4PP. In case of CPD photoproduct (cause little distortion), XPC complex binds to the lesion after recruitment of UV-DDB whereas 6-4PP that distorts the DNA helix to a great extent can be recognized either by interacting with prebound UV-DDB or directly by XPC complex.

After initial steps of damage recognition, the subsequent pathway for both GGR and TCR system is almost similar. The unwinding of DNA double helix at the site of lesion takes place by the components of multi-subunit transcription factor-IIH (TFIIH). TFIIH is a ten-subunit protein complex (Table 4) composed of a core complex (XPB, XPD, p62, p44, p34, p52, p8) and of a cdk activating kinase (CAK) subunit (Mat1, Cdk7, CyclinH) [236]. TFIIH is usually involved in initiation of RNA Pol II transcription, but upon DNA damage can be employed in cell cycle regulation and NER (both in global genome and TC-NER) [228, 235]. Two subunits of TFIIH such as XPB (3′ to 5′ helicase) and XPD (5′ to 3′ helicase) are responsible for opening of DNA double helix around the lesion in an energy (ATP) dependent manner. It has been found that the XPD helicase activity is dispensable for in vitro transcription [262] but seems to play an additional architectural role within the complex by connecting the core TFIIH with the CAK subunit [228, 236]. After opening of DNA double helix by TFIIH, three proteins such as RPA, XPA, and XPG are recruited. The exact order of assembly of these proteins is not clear. Both XPA and RPA can recruit with the DNA lesions in absence of XPG, and similarly XPG can also join the damage sites in the absence of XPA [238, 239]. Moreover, XPA and heterotrimeric replication protein A (RPA; RPA70, 32 and 14) are recruited to confirm the presence of DNA damage and form a more stable preincision complex [185]. RPA was found to be required both for the dual incision and for the repair synthesis steps of NER [263]. It was assumed that RPA binds the nondamaged strand of the opened DNA bubble, thus allowing exact positioning and stimulation of the endonuclease activities of XPG and ERCC1-XPF [246, 247]. XPG, which belongs to flap endonuclease-1 (FEN-1) family of structure-specific endonucleases [264], is not only involved in performing the 3′ incision in NER but also required for stabilizing the fully open DNA bubble structure and to permit the 5′ incision by ERCC1- XPF [265]. Subsequently, the injured part of the DNA is removed by cleaving the damaged strand towards 3′ and 5′ of the lesion by endonuclease XPG and XPF/ERCC1 complex, respectively, generating a 24–32 base oligonucleotide fragment [183]. It has been found that E2F1 plays a direct, non-transcriptional role in DNA repair involving increased recruitment of NER factors to sites of UV-induced DNA damage [266]. Finally the gap is filled by DNA polymerase δ or ε (along with some accessory proteins, like PCNA and RFC) and sealed by DNA ligase. It is assumed that ligase-I is responsible for ligation of remaining nick [267], but very recently it has been reported that mostly ligase III, in cooperation with its partner XRCC1 seals the DNA nicks and ligase I plays a minor role in actively replicating cells, but not in quiescent cells [268]. Recently, several workers have tried to dissect the molecular mechanisms of TC-NER [216, 223, 244, 269]. In spite of the above facts, more investigations are still required to improve our understanding of the GGR and TCR pathways.

Several observations provide evidence that "dark" repair of UV-induced NER is a significant DNA repair mechanisms in plants that is capable of excising dimers, particularly 6-4PPs [30, 270]. Genetic and genomic analysis indicates that plant NER pathway is homologous to that of mammals and fungi and unrelated to the bacterial system [204, 271–273]. Based on the reduction of nuclear CPD frequency, the presence of NER has been reported to occur in several plants such as Glycine max and cultivars of Oryza sativa [274, 275]. Furthermore, CPDs were found to be excised from the nuclear DNA of Daucus carota and Wolffia microscopiaat rates dependent on damage levels and comparable to those in animal cells [205, 275]. A UV-specific endonuclease resembling UvrABC nuclease in activity was partially characterized from spinach [206]. Classical genetic analysis has resulted in the identification of at least four complementation groups required for this repair in Arabidopsis (UVR1, UVR5, UVR7 and UVH1) [276–278] and many more UV-sensitive mutants await further genetic and phenotypic characterization [206]. Moreover, a plant homologue of human NER gene of the endonuclease, ERCC1, has been cloned from Lilium longiflorum which showed a similar role in DNA repair in plants [279].

RECOMBINATIONAL REPAIR

It is one of the widespread mechanisms which efficiently repair double-strand breaks (DSBs) and single-strand gaps in damaged DNA by a series of complex biochemical reactions, as a result of ionizing radiation, UVR, ROS, and chemotherapeutic genotoxic chemicals [5]. The lethal effects of double strand breaks (DSBs) can be conquered by the existence of two independent pathway, such as homologous recombination (HR) and non-homologous end joining (NHEJ). Multiple proteins are required for DSB repair by recombination, which are conserved among all eukaryotes and deficiencies in this repair mechanism can cause hereditary diseases. For instance, mutation of at least one of these repair proteins, called BRCA1 may lead to hereditary breast cancer [280]. Evidence suggests that BRCA1 could be one of the key players in DNA damage response [281]. DSB repair through HR process is an error free pathway, since, it requires an extensive region of sequence homology between the damaged and template strands, whereas NHEJ is an error prone, alternate pathway for the repair of DSBs, essentially joins broken chromosomal ends independent of sequence homology.

Homologous Recombination

Repair of DSBs by HR requires the genes of "RAD52 epistasis group" such as RAD50, RAD51, RAD52, RAD54,RAD55, RAD57, RAD59, MRE11 and XRS2, which were first defined in yeast Saccharomyces cerevisiaemutants. Homologues of most of the genes are highly conserved among all eukaryotes including human [282,283] highlighting these genes for cell survival. Among all, E. coli recA gene and its eukaryotic homologsRAD51s are the best recombination genes [284]. The recA gene encodes a DNA-dependent ATPase that binds to ssDNA and promotes strand invasion and exchange between homologous DNA molecules [285]. Among eukaryotes, the yeast S. cerevisiae and Schizosaccharomyces pombe have four RAD51-like genes (RAD51, DMC1, RAD55/rhp55, and RAD57/rhp57) [286, 287] whereas vertebrate animals and plants have seven types ofRAD51-like genes (RAD51, -51B, -51C, -51D, DMC1, XRCC2, and XRCC3) [288]. The eukaryotic RAD51s play a significant role in HR, maintaining genome integrity in both mitotic as well as meiotic cell cycle [288, 289]. There are 59% identity between S. cerevisiae, human and mouse in the case of Rad51 protein (Rad51p) and 30% identity to RecA protein of bacteria [290], whereas yeast and human proteins are 60% identical in case of Rad52 proteins (Rad52p) [291]. Yeast cell can exhibit both "allelic" [292, 293] as well as "ecotopic recombination" [294, 295] to repair a broken chromosome. Recombinational repair is a significant UV-tolerance mechanism in plants where UV-induced chromosomal rearrangements including homologous intrachromosomal recombination events have been found [296]. In mammalian cells the recombinational pathway of DSBs, seems to be operated in late S- and G2-phase, when DNA molecules are replicated and spatially juxtaposed [297].

The first step of DSB repair via HR is the resection of 5′ ends to produce a 3′ ssDNA overhang by means of an exonuclease (such as RecBCD in E. coli, MRX-complex in S. cerevisiae and MRN-complex in vertebrates). Rad51 (a functional homolog of the E. coli RecA) [298] is the central protein in HR, binds the exposed single-stranded tails forming a nucleoprotein filament and this early step is promoted by a Rad55/Rad57 protein heterodimer [299] by overcoming the inhibitory effects of the heterotrimeric single-stranded DNA binding protein RPA [300]. Recently, it has been reported that a member of the histone H2A family, γ-H2AX protein plays an important role in the recruitment of Rad51 protein in HR in eukaryotes [301]. The Rad51 nucleoprotein filament in association with other repair protein searches the genome for an intact copy of the broken DNA on the sister chromatid to form a heteroduplex joint molecules or D-loop that is matured in to a Holliday junctions (HJs). HJ is then resolved to give crossover products (Figure 13). In E. coli, this HJ is resolved by the positioning of RuvABC resolvasome, however in eukaryotic cell, how this HJ is resolved to give crossover products is not known. In S. cerevisiae, DNA DSB repair by forming noncrossover product has been reported which does not involve the processing of a double HJ structure. Virtually, the joint molecule formation is followed by extension of the incoming strand by DNA polymerases and branch migration, leading to restoration of the genetic information [283, 302].

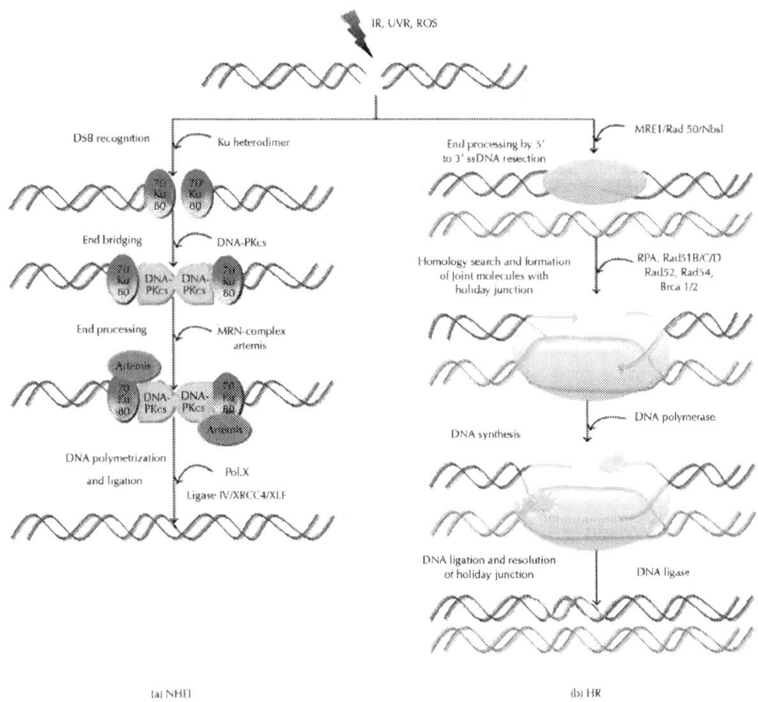

Figure 13: Schematic representation of recombinational repair by (a) non-homologous end joining (NHEJ), and (b) homologous recombination (HR).

Non-Homologous End Joining

When HR is inactivated, an alternate pathway, that is, NHEJ becomes operative for the repair of DSBs [303], that also involves a multi-protein complex and has been found in organisms ranging from a few prokaryotes to mammals. This suggests that this mechanism has been conserved during the course of evolution, although, most of the protein factors, involved in NHEJ were initialy identified in the mammalian cells [304]. Most of the DSBs produced by DNA damaging agents, do not have ligatable termini, hence requires the action of nucleases and DNA polymerases to generate them. The participation of DNA polymerase in the NHEJ pathways is still a matter of debate, however in vitro biochemical analysis in mammals, suggest that DNA polymerase λ (Pol. λ) and/or polymerase μ (Pol. μ) participates in NHEJ process at incompatible DNA ends [80, 305, 306]. The NHEJ process is initiated by the binding of specific protein to the broken ends, which may acts as end bridging factor [307]. It has been shown that Ku complex (a heterodimer of Ku70/Ku80 [≈86]) is a major end binding factor in mammalian cells, possess end bridging activity [308, 309]. The catalytic subunit of DNA protein kinase (DNA-PKcs) is required in mammalian NHEJ to bridge the DNA ends through their protein-protein interactions [310, 311]. Cells lacking functional DNA-PK components are known to have elevated sensitivity toward UV irradiation [312].

Association of DNA-PKcs is followed by the recruitment of other repair proteins (such as ligaseIV/Xrcc4, Artemis, PNK and Polymerase X) to proceed the NHEJ repair [304] (Figure 13). Recently, a third protein, designated as XLF or Cernunnos [313] that has homology to Xrcc4, has been identified and shown to co-associate with the DNA ligaseIV/Xrcc4 complex [314, 315]. Artemis, a member of β-lactamase superfamily, has exonuclease activity. In the presence of DNA-PKcs, Artemis can also function as a endonuclease. It has been found that Artemis dependant DSB rejoining also requires ATM, Mre11-Rad50-Nbs1 (MRN) complex, 53BP1 and H2AX [316–318]. The yeast Hdf1/2 and Dnl4/Lif1 are the functional homology of mammalian Ku and DNA ligaseIV/Xrcc4, respectively. DNA ligase IV is absent in bacteria, however, in Bacillus subtilis, the gene ykou/v has been found that encodes a polypeptide with ligase, primase and nuclease domains. Genetic and biochemical evidence suggest that Mre11-Rad50-Xrs2 (MRX) act as end bridging factor in yeast NHEJ, since DNA-PKcs is absent in them [319, 320].

In case of bacteria, the Ku proteins occur in homodimeric forms and exhibits homology with eukaryotic Ku protein in some extent [321]. Recombinational repair is a significant UV tolerance mechanism in plants where UV-induced chromosomal rearrangements including homologous intrachromosomal recombination events have been found [296]. Very few of the plant genes involved in DSB repair have been identified. The sequence of an Arabidopsis Rad51homologue has been made available [322]. It has been suggested that as in mammals, breaks are repaired by nonhomologous recombination far more frequently than via HR [202, 323].

SOS REPAIR/RESPONSE

The accumulation of massive amount of DNA lesions within the cells under different specific physiological responses [324] may lead to the occurrence of SOS repair system which was well described in E. coli, where the involvement of more than 40 genes have been found [325]. It has been found that the bacterial NER is linked with all DNA damage response through a network of reactions, known as SOS response [326].

The accumulation of DNA lesions may interfere with replication process, prompting cells to stop division, therefore giving time to the cell to repair damaged DNA and proceeds DNA replication process [324]. SOS repair system is initiated by interaction of two crucial proteins the RecA and the LexA repressor which curbs the expression of SOS genes by binding to their promoters [327] (Figure 14). The proteolytic activity of RecA protein inactivates the LexA repressor and induces all the genes to which LexA is associated. A number of genes (or operon) collectively known as din (damage inducible) gene such as uvrA, uvrB, cho (uvrC homolog) and uvrD of E. coli NER take part in SOS response [325, 328]. The SOS response is induced with damage signal but it is highly mutagenic due to engagement of error-prone DNA polymeraseV (UmuC/UmuD$_2$ complex) [329] and DNA polymerase IV [330] in E. coli. Interestingly, it has been found that DNA polymerase IV (dinB) is also involved in translesion synthesis in E. coli [331]. Majchrzak et al. [324] examined the effects of SOS response on genome stability of trinucleotide repeat sequences (TRS) in E. coli and observed that SOS response genes destabilized the TRS tracts and also altered the superhelical density of the plasmids. Recently, the genesimuA and imuB have been described that induce SOS mutagenesis in Caulobacter crescentus, but absent in E. coli [332]. However, the number of genes responsible for SOS repair system has still to be investigated.

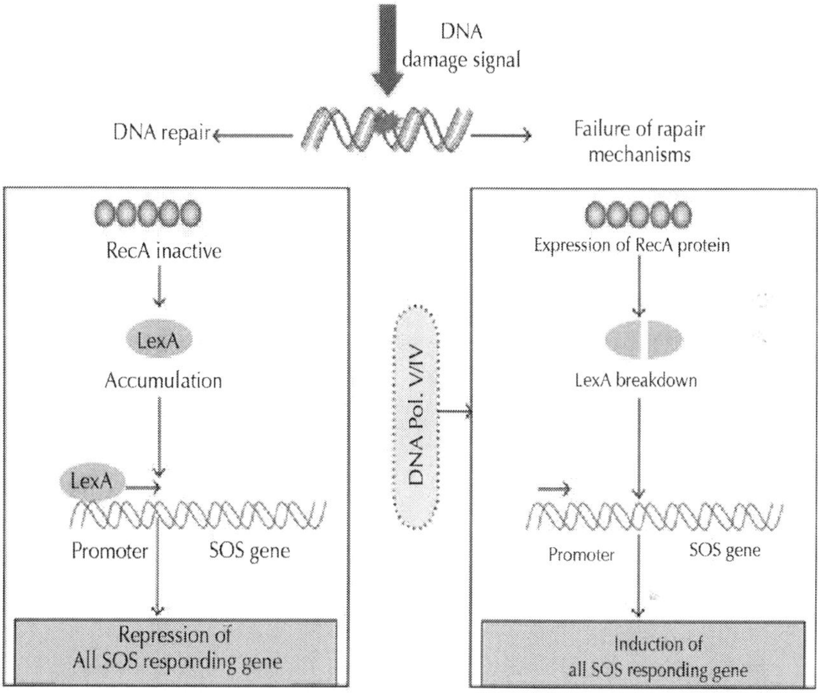

Figure 14: SOS response: As a result of massive DNA damage and failure of all possible repair mechanisms, RecA proteins is expressed, which activate the auto breakdown of LexA proteins, allowing the induction of all SOS responding genes. The pathway of SOS response is reversed when damages are repaired through the damage specific mechanisms. Here the inactivation of RecA protein allows the accumulation of LexA, which bind to SOS promoters and repress all SOS responding genes. SOS response is highly mutagenic due to involvement of DNA polymerase V/IV.

CELL-CYCLE CHECKPOINT ACTIVATION

In response to diverse genotoxic stresses such as UV radiation, IR, chemicals used in medical therapy, by-products of intra-cellular metabolism, several protective mechanisms including processes of DNA repair, Cell-cycle checkpoint arrest and apoptosis (programmed cell death) develop within the organisms to secure the genomic integrity. DNA damage such as altered DNA bases, abasic site, DNA lesions (CPDs/6-4PPs), strand breaks (DSBs/SSBs) may stop the progression of cell-cycle temporarily to give opportunities to the cell for DNA repair before replication or segregation of the affected chromosome [333], or may induce an apoptotic program to eliminate the damaged cells to avoid their carcinogenic potential [334]. ROS may induce several types of DNA lesions such as DSBs and SSBs, DNA-DNA and DNA-protein cross links and base modifications [335]. It has been shown that most of the chromosomal translocations produce via aberrant processing of a DNA DSBs. Hence forth, we discuss the DSBs/SSBs-induced cell-cycle checkpoint arrest in eukaryotes mainly in mammals that play a critical role in preventing chromosomal instability. Regulation of cell-cycle checkpoint proceeds through a network of damage sensors, signal transducers, mediators, and various effector proteins [183]. Phosphatidylinositol-3 (PI3)-kinase related kinases (PIKKs) ATM (ataxia telangiectasia mutated) protein, ATR (ATM and Rad3 related) protein, and DNA-PK, with effector proteins mediated cell-cycle checkpoint arrest (at G1/S, G2/M, and intra S-phase), DNA repair and cell death have been observed in mammalian cells [336]. It has been shown that ATM and DNA-PK are activated by the presence of DSBs where as activation of ATR takes place by single strand regions of DNA [337, 338]. Activation of both ATM and ATR results in phosphorylation of Chk2 and Chk1 respectively which transfer the DNA damage signal to the cell-division cycle proteins Cdc25(A-C). Phosphorylation of Cdc25 by Chk1/Chk2 leads to its ubiquitin mediated degradation that results in G1 and S-phase arrest (Figure 15). The prevailing evidence suggests that damage response mediated activation of ATM/ATR either directly or via Chk2 phosphorylates p53, which transcriptionally activates the Cdk inhibitor, p21, which arrest G1/S cell-cycle checkpoint [336]. Recently, it has been reported that DNA damage caused by UV radiation or ROS such as hydroxyl (˙OH) free-radical results in ATM mediated phosphorylation of BID protein that induce cell-cycle arrest in S-phase [339, 340]. The occurrence of DNA damage response in G2-phase, leads to checkpoint mediator (claspin) dependent activation of Chk1/2, followed by $SCF^{\beta TrcP}$ mediated degradation of CDK-activating phosphatase Cdc25A [341, 342], that results in arrest of multiple cell-cycle transition including the G2 checkpoint [343, 344]. The ubiquitin mediated destruction of claspin and WEE1 (both proteins have conserved β-TrcP phosphodegrons) eliminates the essential coactivator of Chk1 and CDK inhibitor respectively allowing reaccumulation of Cdc25A followed by Cdc25B and C, which results in activation of cyclin-Cdk (cyclinB-Cdk1) complex [345]. Under normal conditions, this latter complex being active promotes G2/M transition and upon inactivation due to DNA damage, blocks the G2 cell-cycle and unlike the G1/S checkpoint this arrest seems to be partly p53/p21 independent [183]. The intermediate component of this pathway is β-TrcP (an adaptor protein) that links both WEE1 [346] and claspin [347–349] with the SCF ubiquitin ligase complex and this $SCF^{\beta TrcP}$ acts as trigger of checkpoint initiation where recognition of phosphodegron β-TrcP is exposed after Chk1 mediated Phosphorylation of Cdc25A [350], as well as checkpoint recovery which is linked with Plk1 mediated phosphorylation of claspin and WEE1 [347, 348]. Although, the exact mechanism(s) regarding the reactivation of Plk1 during checkpoint recovery is still in dispute. The malfunctioning of cell-cycle checkpoint as a result of chronic damage and/or defects in DNA damage response (DDR) components such as p53, p21, ATM, Chk2, BRCA1/2 tumor suppressors, may induce several types of human disorder at the expence of enhanced genomic instability [351].

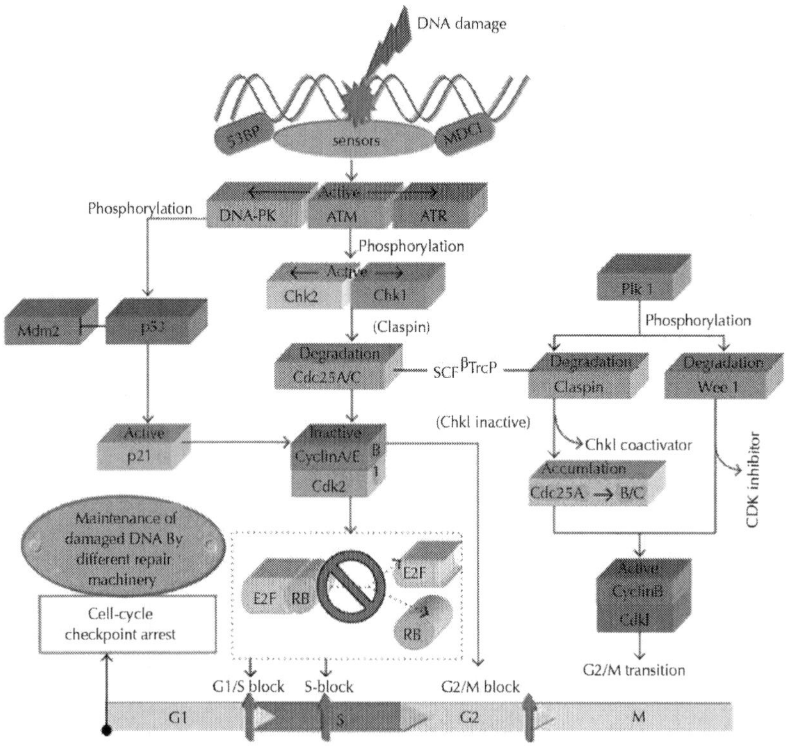

Figure 15: Schematic illustration of DNA damage-induced cell-cycle checkpoint activation (for details, see text).

In addition to the above mentioned repair mechanisms several other repair machineries such as mutagenic repair (or lesion bypass) and programmed cell death (PCD) or apoptosis may become effective for the recovery of genome against constant attack of numerous genotoxins. UV-radiation, ionizing radiation and various chemicals are responsible for most of the mutagenesis due to a process of translesion synthesis in which a polymerase or replicative assembly encounters a noncoding or miscoding lesion, inserts an incorrect nucleotide opposite the lesion and then continues elongation [5]. It has been reported that translesion synthesis past a CPD, facilitated by pol. η, with the insertion of adenines opposite both bases of a CPD [352], where as 6-4TT may lead to a G insertion. In Saccharomyces cerevisiae, pol. η and pol. ζ (consisting of Rev3 and Rev7 proteins) has been reported to replicate across a CPD

[353]. Polymerase η (pol. η) can also replicate across a basic sites, AAF (acetylaminofluorene), guanine adducts and cis-platinated guanines [352].

When the repair mechanisms are unsuccessful, it may cause cellular senescence (permanent cell cycle arrest), oncogenesis or apoptosis [354]. Apoptosis plays an essential role in survival of the organisms by preventing the multiplication of mutated chromosomes, normal embryonic development, elimination of indisposed cells and maintenance of cell homeostasis.

ACKNOWLEDGMENTS

R. P. Rastogi acknowledges University Grants Commission, New Delhi, India, for financial support in the form of a fellowship. Works related to UV-B radiation effects are partly supported by the project grant sanctioned by Department of Science and Technology, Govt. of India, New Delhi (SR/SO/PS-49/09).

REFERENCES

1. D. Lubin and E. H. Jensen, "Effects of clouds and stratospheric ozone depletion on ultraviolet radiation trends," Nature, vol. 377, no. 6551, pp. 710–713, 1995.
2. R. P. Sinha, H. D. Kumar, A. Kumar, and D. P. Hader, "Effects of UV-B irradiation on growth, survival, pigmentation and nitrogen metabolism enzymes in cyanobacteria," Acta Protozoologica, vol. 34, no. 3, pp. 187–192, 1995.
3. R. P. Sinha, R. P. Rastogi, N. K. Ambasht, and D.-P. Häder, "Life of wetland cyanobacteria under enhancing solar UV-B radiation," Proceedings of the National Academy of Sciences, India B, vol. 78, pp. 53–65, 2008.
4. D.-P. Häder and R. P. Sinha, "Solar ultraviolet radiation-induced DNA damage in aquatic organisms: potential environmental impact," Mutation Research, vol. 571, no. 1-2, pp. 221–233, 2005.

5. E. C. Friedberg, G. C. Walker, W. Siede, R. D. Wood, R. A. Schultz, and T. Ellenberger, DNA Repair and Mutagenesis, ASM Press, Washington, DC, USA, 2006.
6. D.-P. Häder, H. D. Kumar, R. C. Smith, and R. C. Worrest, "Effects of solar UV radiation on aquatic ecosystems and interactions with climate change," Photochemical and Photobiological Sciences, vol. 6, no. 3, pp. 267–285, 2007.
7. M. Norval, A. P. Cullen, F. R. De Gruijl et al., "The effects on human health from stratospheric ozone depletion and its interactions with climate change," Photochemical and Photobiological Sciences, vol. 6, no. 3, pp. 232–251, 2007.
8. K. R. Solomon, "Effects of ozone depletion and UV-B radiation on humans and the environment,"Atmosphere—Ocean, vol. 46, no. 1, pp. 185–202, 2008.
9. M. Zeeshan and S. M. Prasad, "Differential response of growth, photosynthesis, antioxidant enzymes and lipid peroxidation to UV-B radiation in three cyanobacteria," South African Journal of Botany, vol. 75, no. 3, pp. 466–474, 2009.
10. M. Llabrés, S. Agustí, P. Alonso-Laita, and G. J. Herndl, "Synechococcus and Prochlorococcus cell death induced by UV radiation and the penetration of lethal UVR in the Mediterranean Sea," Marine Ecology Progress Series, vol. 399, pp. 27–37, 2010
11. A. Tabazadeh, M. L. Santee, M. Y. Danilin, et al., "Quantifying denitrification and its effect on ozone recovery," Science, vol. 288, no. 5470, pp. 1407–1411, 2000.
12. R. L. McKenzie, L. O. Björn, A. Bais, and M. Ilyasd, "Changes in biologically active ultraviolet radiation reaching the Earth's surface," Photochemical and Photobiological Sciences, vol. 2, no. 1, pp. 5–15, 2003
13. R. G. Alscher, J. L. Donahue, and C. L. Cramer, "Reactive oxygen species and antioxidants: relationships in green cells," Physiologia Plantarum, vol. 100, no. 2, pp. 224–233, 1997
14. L. Chalker-Scott, "Survival and sex ratios of the intertidal copepod, Tigriopus californicus, following ultraviolet-B (290–320 nm) radiation exposure," Marine Biology, vol. 123, no. 4, pp. 799–804, 1995.

15. M. J. Peak and J. G. Peak, "Single-strand breaks induced in Bacillus subtilis DNA by ultraviolet light: action spectrum and properties," Photochemistry and Photobiology, vol. 35, no. 5, pp. 675–680, 1982.
16. M. J. Peak, J. G. Peak, M. P. Moehring, and R. B. Webb, "Ultraviolet action spectra for DNA dimer induction, lethality, and mutagenesis in Escherichia coli with emphasis on the UVB region,"Photochemistry and Photobiology, vol. 40, no. 5, pp. 613–620, 1984.
17. R. P. Sinha, M. Dautz, and D.-P. Häder, "A simple and efficient method for the quantitative analysis of thymine dimers in cyanobacteria, phytoplankton and macroalgae," Acta Protozoologica, vol. 40, no. 3, pp. 187–195, 2001.
18. A. G. J. Buma, M. K. De Boer, and P. Boelen, "Depth distributions of DNA damage in antarctic marine phyto and bacterioplankton exposed to summertime UV radiation," Journal of Phycology, vol. 37, no. 2, pp. 200–208, 2001
19. H. Pakker, C. A. C. Beekman, and A. M. Breeman, "Efficient photoreactivation of UVBR-induced DNA damage in the sublittoral macroalga Rhodymenia pseudopalmata (Rhodophyta)," European Journal of Phycology, vol. 35, no. 2, pp. 109–114, 2000
20. F. E. Quaite, B. M. Sutherland, and J. C. Sutherland, "Quantitation of pyrimidine dimers in DNA from UVB-irradiated alfalfa (Medicago sativa L.) seedlings," Applied and Theoretical Electrophoresis, vol. 2, no. 6, pp. 171–175, 1992.
21. B. Stein, H. J. Rahmsdorf, A. Steffen, M. Litfin, and P. Herrlich, "UV-induced DNA damage is an intermediate step in UV-induced expression of human immunodeficiency virus type 1, collagenase, c-fos, and metallothionein," Molecular and Cellular Biology, vol. 9, no. 11, pp. 5169–5181, 1989.
22. R. P. Sinha and D.-P. Häder, "UV-induced DNA damage and repair: a review," Photochemical and Photobiological Sciences, vol. 1, no. 4, pp. 225–236, 2002
23. K. M. D. Lima-Bessa, M. G. Armelini, V. Chiganças et al., "CPDs and 6–4PPs play different roles in UV-induced cell death in normal and NER-deficient human cells," DNA Repair, vol. 7, no. 2, pp. 303–312, 2008

24. K. McAteer, Y. Jing, J. Kao, J.-S. Taylor, and M. A. Kennedy, "Solution-state structure of a DNA dodecamer duplex containing a cis-syn thymine cyclobutane dimer, the major UV photoproduct of DNA," Journal of Molecular Biology, vol. 282, no. 5, pp. 1013–1032, 1998
25. J.-H. Lee, G.-S. Hwang, and B.-S. Choi, "Solution structure of a DNA decamer duplex containing the stable 3′ T·G base pair of the pyrimidine(6–4)pyrimidone photoproduct [(6–4) adduct]: implications for the highly specific 3′ T → C transition of the (6–4) adduct," Proceedings of the National Academy of Sciences of the United States of America, vol. 96, no. 12, pp. 6632–6636, 1999.
26. J.-H. Lee, S.-H. Bae, and B.-S. Choi, "The Dewar photoproduct of thymidylyl(3′→5′)-thymidine (Dewar product) exhibits mutagenic behavior in accordance with the "A rule"," Proceedings of the National Academy of Sciences of the United States of America, vol. 97, no. 9, pp. 4591–4596, 2000.
27. DeLano Scientific LLC, 400 Oyster Point Blvd., Suite 213, South San Francisco, CA 94080-1918, USA.http://www.pymol.org/.
28. R. P. Sinha, M. Krywult, and D.-P. Häder, "Effects of ultraviolet, monochromatic and PAR waveband on nitrate reductase activity and pigmentation in a rice field cyanobacterium, Anabaena sp," Acta Hydrobiologica, vol. 2, pp. 105–112, 1998.
29. R. P. Sinha, S. P. Singh, and D.-P. Häder, "Database on mycosporines and mycosporine-like amino acids (MAAs) in fungi, cyanobacteria, macroalgae, phytoplankton and animals," Journal of Photochemistry and Photobiology B, vol. 89, no. 1, pp. 29–35, 2007
30. A. B. Britt, "Repair of DNA damage induced by solar UV," Photosynthesis Research, vol. 81, no. 2, pp. 105–112, 2004
31. R. P. Rastogi, Richa, R. P. Sinha, S. P. Singh, and D.-P. Häder, "Photoprotective compounds from marine organisms," Journal of Industrial Microbiology and Biotechnology, vol. 37, no. 6, pp. 537–558, 2010
32. R. P. Rastogi, Richa, S. P. Singh, D.-P. Häder, and R. P. Sinha, "Mycosporine-like amino acids profile and their activity under PAR and UVR in a hot-spring cyanobacterium Scytonema sp. HKAR-3,"Australian Journal of Botany, vol. 58, no. 4, pp. 286–293, 2010

33. Z. Xie, Y. Wang, Y. Liu, and Y. Liu, "Ultraviolet-B exposure induces photo-oxidative damage and subsequent repair strategies in a desert cyanobacterium Microcoleus vaginatus Gom," European Journal of Soil Biology, vol. 45, no. 4, pp. 377–382, 2009
34. H. Wiseman and B. Halliwell, "Damage to DNA by reactive oxygen and nitrogen species: role in inflammatory disease and progression to cancer," Biochemical Journal, vol. 313, no. 1, pp. 17–29, 1996.
35. J.-S. Taylor, "DNA, sunlight, and skin cancer," Journal of Chemical Education, vol. 67, no. 10, pp. 833–841, 1990.
36. T. Lindahl, "Instability and decay of the primary structure of DNA," Nature, vol. 362, no. 6422, pp. 709–715, 1993
37. M. Valko, C. J. Rhodes, J. Moncol, M. Izakovic, and M. Mazur, "Free radicals, metals and antioxidants in oxidative stress-induced cancer," Chemico-Biological Interactions, vol. 160, no. 1, pp. 1–40, 2006
38. B. Halliwell and J. Gutteridge, Free Radicals in Biology and Medicine, Oxford University Press, Oxford, UK, 4th edition, 2007.
39. M. Dizdaroglu, "Oxidative damage to DNA in mammalian chromatin," Mutation Research, vol. 275, no. 3–6, pp. 331–342, 1992
40. M. Valko, D. Leibfritz, J. Moncol, M. T. D. Cronin, M. Mazur, and J. Telser, "Free radicals and antioxidants in normal physiological functions and human disease," International Journal of Biochemistry and Cell Biology, vol. 39, no. 1, pp. 44–84, 2007
41. Y.-H. You, P. E. Szabo, and G. P. Pfeifer, "Cyclobutane pyrimidine dimers form preferentially at the major p^{53} mutational hotspot in UVB-induced mouse skin tumors," Carcinogenesis, vol. 21, no. 11, pp. 2113–2117, 2000.
42. J.-L. Ravanat, T. Douki, and J. Cadet, "Direct and indirect effects of UV radiation on DNA and its components," Journal of Photochemistry and Photobiology B, vol. 63, no. 1–3, pp. 88–102, 2001
43. J. Cadet, E. Sage, and T. Douki, "Ultraviolet radiation-mediated damage to cellular DNA," Mutation Research, vol. 571, no. 1-2, pp. 3–17, 2005

44. D. I. Pattison and M. J. Davies, "Actions of ultraviolet light on cellular structures," in Cancer: Cell Structures, Carcinogens and Genomic Instability, L. P. Bignold, Ed., Birkhäuser, Switzerland, 2006.

45. J. S. Taylor, H. F. Lu, and J. J. Kotyk, "Quantitative conversion of the (6–4) photoproduct of TpdC to its Dewar valence isomer upon exposure to simulated sunlight," Photochemistry and Photobiology, vol. 51, no. 2, pp. 161–167, 1990.

46. S. Courdavault, C. Baudouin, M. Charveron et al., "Repair of the three main types of bipyrimidine DNA photoproducts in human keratinocytes exposed to UVB and UVA radiations," DNA Repair, vol. 4, no. 7, pp. 836–844, 2005

47. T. Carell and R. Epple, "Repair of UV light induced DNA lesions: a comparative study with model compounds," European Journal of Organic Chemistry, no. 7, pp. 1245–1258, 1998.

48. T. Douki, A. Reynaud-Angelin, J. Cadet, and E. Sage, "Bipyrimidine photoproducts rather than oxidative lesions are the main type of DNA damage involved in the genotoxic effect of solar UVA radiation," Biochemistry, vol. 42, no. 30, pp. 9221–9226, 2003

49. J. E. Donnellan Jr. and R. B. Setlow, "Thymine photoproducts but not thymine dinners found in ultraviolet-irradiated bacterial spores," Science, vol. 149, no. 3681, pp. 308–310, 1965.

50. M. M. Becker and Z. Wang, "Origin of ultraviolet damage in DNA," Journal of Molecular Biology, vol. 210, no. 3, pp. 429–438, 1989

51. V. Lyamichev, "Unusual conformation of (dA)n.(dT)n-tracts as revealed by cyclobutane thymine-thymine dimer formation," Nucleic Acids Research, vol. 19, no. 16, pp. 4491–4496, 1991.

52. J. R. Pehrson and L. H. Cohen, "Effects of DNA looping on pyrimidine dimer formation," Nucleic Acids Research, vol. 20, no. 6, pp. 1321–1324, 1992.

53. A. Aboussekhra and F. Thoma, "TATA-binding protein promotes the selective formation of UV-induced (6–4)-photoproducts and modulates DNA repair in the TATA box," EMBO Journal, vol. 18, no. 2, pp. 433–443, 1999

54. T. Douki and J. Cadet, "Individual determination of the yield of the main UV-induced dimeric pyrimidine photoproducts in DNA

suggests a high mutagenicity of CC photolesions," Biochemistry, vol. 40, no. 8, pp. 2495–2501, 2001

55. A. S. Balajee andV. A. Bohr, "Genomic heterogeneity of nucleotide excision repair," Gene, vol. 250, no. 1-2, pp. 15–30, 2000
56. J. M. Gale, K. A. Nissen, and M. J. Smerdon, "UV-induced formation of pyrimidine dimers in nucleosome core DNA is strongly modulated with a period of 10.3 bases," Proceedings of the National Academy of Sciences of the United States of America, vol. 84, no. 19, pp. 6644–6648, 1987.
57. R. M. Tyrrell, "Induction of pyrimidine dimers in bacterial DNA by 365 nm radiation," Photochemistry and Photobiology, vol. 17, no. 1, pp. 69–73, 1973.
58. E. Kvam and R. M. Tyrrell, "Induction of oxidative DNA base damage in human skin cells by UV and near visible radiation," Carcinogenesis, vol. 18, no. 12, pp. 2379–2384, 1997
59. D. Perdiz, P. Gróf, M. Mezzina, O. Nikaido, E. Moustacchi, and E. Sage, "Distribution and repair of bipyrimidine photoproducts in solar UV-irradiated mammalian cells: possible role of dewar photoproducts in solar mutagenesis," Journal of Biological Chemistry, vol. 275, no. 35, pp. 26732–26742, 2000
60. S. Courdavault, C. Baudouin, M. Charveron, A. Favier, J. Cadet, and T. Douki, "Larger yield of cyclobutane dimers than 8-oxo-7,8-dihydroguanine in the DNA of UVA-irradiated human skin cells,"Mutation Research, vol. 556, no. 1-2, pp. 135–142, 2004
61. D. G. T. Su, J.-S. A. Taylor, and M. L. Gross, "A new photoproduct of 5-methylcytosine and adenine characterized by high-performance liquid chromatography and mass spectrometry," Chemical Research in Toxicology, vol. 23, no. 3, pp. 474–479, 2010
62. K. L. Wierzchowski and D. Shugar, "Photochemistry of cytosine nucleosides and nucleotides, II," Acta Biochimica Polonica, vol. 8, pp. 219–234, 1961.
63. R. J. Boorstein, T. P. Hilbert, R. P. Cunningham, and G. W. Teebor, "Formation and stability of repairable pyrimidine photohydrates in DNA," Biochemistry, vol. 29, no. 46, pp. 10455–10460, 1990.
64. N. J. Duker and P. E. Gallagher, "Purine photoproducts," Photochemistry and Photobiology, vol. 48, no. 1, pp. 35–39, 1988.

65. S. Kumar, N. D. Sharma, R. J. Davies, D. W. Phillipson, and J. A. McCloskey, "The isolation and characterisation of a new type of dimeric adenine photoproduct in UV-irradiated deoxyadenylates,"Nucleic Acids Research, vol. 15, no. 3, pp. 1199–1216, 1987.
66. T. M. G. Koning, R. J. H. Davies, and R. Kaptein, "The solution structure of the intramolecular photoproduct of d(TpA) derived with the use of NMR and a combination of distance geometry and molecular dynamics," Nucleic Acids Research, vol. 18, no. 2, pp. 277–284, 1990.
67. S. N. Bose, R. J. H. Davies, S. K. Sethi, and J. A. McCloskey, "Formation of an adenine-thymine photoadduct in the deoxydinucleoside monophosphate d(TpA) and in DNA," Science, vol. 220, no. 4598, pp. 723–725, 1983.
68. S. N. Bose and R. J. Davies, "The photoreactivity of T-A sequences in oligodeoxyribonucleotides and DNA," Nucleic Acids Research, vol. 12, no. 20, pp. 7903–7914, 1984.
69. P. H. Clingen, R. Jeremy, and H. Davies, "Quantum yields of adenine photodimerization in poly(deoxyadenylic acid) and DNA," Journal of Photochemistry and Photobiology B, vol. 38, no. 1, pp. 81–87, 1997
70. D. Porschke, "A specific photoreaction in polydeoxyadenylic acid," Proceedings of the National Academy of Sciences of the United States of America, vol. 70, no. 9, pp. 2683–2686, 1973.
71. N. D. Sharma and R. J. H. Davies, "Extent of formation of a dimeric adenine photoproduct in polynucleotides and DNA," Journal of Photochemistry and Photobiology B, vol. 3, no. 2, pp. 247–258, 1989.
72. R. O. Rahn, "Search for an adenine photoproduct in DNA," Nucleic Acids Research, vol. 3, no. 4, pp. 879–890, 1976.
73. P. W. Doetsch, T. H. Zastawny, A. M. Martin, and M. Dizdaroglu, "Monomeric base damage products from adenine, guanine and thymine induced by exposure of DNA to ultraviolet radiation," Biochemistry, vol. 34, pp. 737–742, 1995.
74. D. B. Hall, R. E. Holmlin, and J. K. Barton, "Oxidative DNA damage through long-range electron transfer," Nature, vol. 382, no. 6593, pp. 731–735, 1996

75. M. Lukin and C. de los Santos, "NMR structures of damaged DNA," Chemical Reviews, vol. 106, no. 2, pp. 607–686, 2006
76. J.-H. Lee, G.-S. Hwang, J.-K. Kim, and B.-S. Choi, "The solution structure of DNA decamer duplex containing the Dewar product of thymidylyl(3′ → 5′)thymidine by NMR and full relaxation matrix refinement," FEBS Letters, vol. 428, no. 3, pp. 269–274, 1998
77. S. N. Rao, J. W. Keepers, and P. Kollman, "The structure of d(CGCGAAT[ ]TCGCG)-d(CGCGAATTCGCG): the incorporation of a thymine photodimer into a B-DNA helix," Nucleic Acids Research, vol. 12, no. 11, pp. 4789–4807, 1984.
78. F. E. Hruska, D. J. Wood, K. K. Ogilvie, and J. L. Charlton, "A proton magnetic resonance study of the ultraviolet photoproduct of d(TpT) in aqueous solution," Canadian Journal of Chemistry, vol. 53, pp. 1193–1203, 1975.
79. L.-S. Kan, L. Voituriez, and J. Cadet, "Nuclear magnetic resonance studies of Cis-Syn, Trans-Syn, and 6–4 photodimers of thymidylyl(3′–5′)thymidine monophosphate and Cis-Syn photodimers of thymidylyl(3′–5′)thymidine cyanoethyl phosphotriester," Biochemistry, vol. 27, no. 15, pp. 5796–5803, 1988.
80. J.-H. Lee, C.-J. Park, J.-S. Shin, T. Ikegami, H. Akutsu, and B.-S. Choi, "NMR structure of the DNA decamer duplex containing double T.G mismatches of cis-syn cyclobutane pyrimidine dimer: implications for DNA damage recognition by the XPC-hHR23B complex," Nucleic Acids Research, vol. 32, no. 8, pp. 2474–2481, 2004
81. C. Suquet and M. J. Smerdon, "UV damage to DNA strongly influences its rotational setting on the histone surface of reconstituted nucleosomes," Journal of Biological Chemistry, vol. 268, no. 32, pp. 23755–23757, 1993.
82. A. E. Rumora, K. M. Kolodziejczak, A. M. Wagner, and M. E. Núñez, "Thymine dimer-induced structural changes to the DNA duplex examined with reactive probes," Biochemistry, vol. 47, no. 49, pp. 13026–13035, 2008
83. J.-K. Kim and B.-S. Choi, "The solution structure of DNA duplex-decamer containing the (6–4) photoproduct of thymidylyl(3′ → 5′) thymidine by NMR and relaxation matrix refinement," European Journal of Biochemistry, vol. 228, no. 3, pp. 849–854, 1995

84. G.-S. Hwang, J.-K. Kim, and B.-S. Choi, "NMR structural studies of DNA decamer duplex containing the Dewar photoproduct of thymidylyl(3′ → 5′)thymidine: conformational changes of the oligonucleotide duplex by photoconversion of a (6–4) adduct to its Dewar valence isomer," European Journal of Biochemistry, vol. 235, no. 1-2, pp. 359–365, 1996.

85. D. A. Pearlman, S. R. Holbrook, D. H. Pirkle, and S.-H. Kim, "Molecular models for DNA damaged by photoreaction," Science, vol. 227, no. 4692, pp. 1304–1308, 1985.

86. T. V. Wang and K. C. Smith, "Postreplication repair in ultraviolet-irradiated human fibroblasts: formation and repair of DNA double-strand breaks," Carcinogenesis, vol. 7, no. 3, pp. 389–392, 1986.

87. T. A. Slieman and W. L. Nicholson, "Artificial and solar UV radiation induces strand breaks and cyclobutane pyrimidine dimers in Bacillus subtilis spore DNA," Applied and Environmental Microbiology, vol. 66, no. 1, pp. 199–205, 2000.

88. C. Baumstark-Khan, U. Hentschel, Y. Nikandrova, J. Krug, and G. Horneck, "Fluorometric analysis of DNA unwinding (FADU) as a method for detecting repair-induced DNA strand breaks in UV-irradiated mammalian cells," Photochemistry and Photobiology, vol. 72, no. 4, pp. 477–484, 2000

89. R. P. Rastogi, S. P. Singh, D.-P. Häder, and R. P. Sinha, "Detection of reactive oxygen species (ROS) by the oxidant-sensing probe 2′,7′-dichlorodihydrofluorescein diacetate in the cyanobacterium Anabaena variabilis PCC 7937," Biochemical and Biophysical Research Communications, vol. 397, no. 3, pp. 603–607, 2010

90. C. L. Limoli, E. Giedzinski, W. M. Bonner, and J. E. Cleaver, "UV-induced replication arrest in the xeroderma pigmentosum variant leads to DNA doublestrand breaks, γ-H2AX formation, and Mre11 relocalization," Proceedings of the National Academy of Sciences of the United States of America, vol. 99, no. 1, pp. 233–238, 2002

91. L. F. Z. Batista, B. Kaina, R. Meneghini, and C. F. M. Menck, "How DNA lesions are turned into powerful killing structures: insights from UV-induced apoptosis," Mutation Research, vol. 681, no. 2-3, pp. 197–208, 2009

92. T. R. Dunkern and B. Kaina, "Cell proliferation and DNA breaks are involved in ultraviolet light-induced apoptosis in nucleotide excision repair-deficient Chinese hamster cells," Molecular Biology of the Cell, vol. 13, no. 1, pp. 348–361, 2002
93. W. P. Roos and B. Kaina, "DNA damage-induced cell death by apoptosis," Trends in Molecular Medicine, vol. 12, no. 9, pp. 440–450, 2006
94. M. M. Vilenchik and A. G. Knudson, "Endogenous DNA double-strand breaks: production, fidelity of repair, and induction of cancer," Proceedings of the National Academy of Sciences of the United States of America, vol. 100, no. 22, pp. 12871–12876, 2003
95. A. Takahashi and T. Ohnishi, "Does γH2AX foci formation depend on the presence of DNA double strand breaks?" Cancer Letters, vol. 229, no. 2, pp. 171–179, 2005
96. G. Wang, L. M. Hallberg, E. Saphier, and E. W. Englander, "Short interspersed DNA element-mediated detection of UVB-induced DNA damage and repair in the mouse genome, in vitro, and in vivo in skin,"Mutation Research, vol. 433, no. 3, pp. 147–157, 1999
97. G. Wang, L. M. Hallberg, E. Saphier, and E. W. Englander, "Pyrimidine (6–4) pyrimidone photoproduct mapping after sub-lethal UVC doses: nucleotide resolution using terminal transferase-dependent PCR,"Photochemistry and Photobiology, vol. 82, pp. 1370–1376, 2006.
98. A. Kumar, M. B. Tyagi, and P. N. Jha, "Evidences showing ultraviolet-B radiation-induced damage of DNA in cyanobacteria and its detection by PCR assay," Biochemical and Biophysical Research Communications, vol. 318, no. 4, pp. 1025–1030, 2004
99. A. Karakoula, M. D. Evans, I. D. Podmore, P. E. Hutchinson, J. Lunec, and M. S. Cooke, "Quantification of UVR-induced DNA damage: global- versus gene-specific levels of thymine dimers," Journal of Immunological Methods, vol. 277, no. 1-2, pp. 27–37, 2003
100. P. L. Olive, J. P. Banath, and R. E. Durand, "Heterogeneity in radiation-induced DNA damage and repair in tumor and normal cells measured using the ‹comet› assay," Radiation Research, vol. 122, no. 1, pp. 86–94, 1990.

101. N. Morley, A. Rapp, H. Dittmar et al., "UVA-induced apoptosis studied by the new apo/necro-Comet-assay which distinguishes viable, apoptotic and necrotic cells," Mutagenesis, vol. 21, no. 2, pp. 105–114, 2006
102. J. L. R. Roti and W. D. Wright, "Visualization of DNA loops in nucleoids from HeLa cells: assays for DNA damage and repair," Cytometry, vol. 8, no. 5, pp. 461–467, 1987.
103. P. Sestili, C. Martinelli, and V. Stocchi, "The fast halo assay: an improved method to quantify genomic DNA strand breakage at the single-cell level," Mutation Research, vol. 607, no. 2, pp. 205–214, 2006
104. J. Chen, K. Jin, M. Chen et al., "Early detection of DNA strand breaks in the brain after transient focal ischemia: implications for the role of DNA damage in apoptosis and neuronal cell death," Journal of Neurochemistry, vol. 69, no. 1, pp. 232–245, 1997.
105. C. Stecco, A. Porzionato, V. Macchi et al., "Detection of apoptosis in human brainstem by TUNEL assay," Italian Journal of Anatomy and Embryology, vol. 110, no. 4, pp. 255–260, 2005.
106. F. Xu, J. M. Vlak, A. P .M. Eker, and M. M. van Oers, "DNA photolyases of Chrysodeixis chalcitesnucleopolyhedrovirus target to the nucleus and interact with chromosomes and mitotic spindle structures," Journal of General Virology, vol. 91, pp. 907–914, 2010.
107. S. K. Murthy and D. J. Demetrick, "New approaches to fluorescence in situ hybridization," Methods in Molecular Biology, vol. 319, pp. 237–259, 2006.
108. S. Kumari, R. P. Rastogi, K. L. Singh, S. P. Singh, and R. P. Sinha, "DNA damage: detection strategies,"EXCLI Journal, vol. 7, pp. 44–62, 2008.
109. G. Koopman, C. P. Reutelingsperger, G. A. Kuiiten, R. M. Keehnen, and S. T. Pals, "Annexin V for flow cytometric detection of phosphotidylserine expression on B cells undergoing apoptosis," Blood, vol. 4, pp. 1415–1420, 1994.
110. J. W. Bickham, "Flow cytometry as a technique to monitor the effects of environmental genotoxins on wild life populations," in In Situ Evaluation of Biological Hazards of Environmental Pollutants, S. S. Sandhu, W. R. Lower, F. J. de Serres, W. A. Suk,

and R. R. Tice, Eds., pp. 97–108, Plenum Press, New York, NY, USA, 1990.

111. H. Watanabe, Y. Murata, M. Miura, M. Hasegawa, T. Kawamoto, and H. Shibuya, "In-vivo visualization of radiation-induced apoptosis using 125I-annexin V," Nuclear Medicine Communications, vol. 27, no. 1, pp. 81–89, 2006

112. M. Klisch, R. P. Sinha, E. W. Helbling, and D.-P. Häder, "Induction of thymine dimers by solar radiation in natural freshwater phytoplankton assemblages in Patagonia, Argentina," Aquatic Sciences, vol. 67, no. 1, pp. 72–78, 2005.

113. S. Otero, E. Núñez-Olivera, J. Martínez-Abaigar, R. Tomás, M. Arróniz-Crespo, and N. Beaucourt, "Effects of cadmium and enhanced UV radiation on the physiology and the concentration of UV-absorbing compounds of the aquatic liverwort Jungermannia exsertifolia subsp. cordifolia,"Photochemical and Photobiological Sciences, vol. 5, no. 8, pp. 760–769, 2006

114. D. L. Mitchell, C. A. Haipek, and J. M. Clarkson, "(6–4) photoproducts are removed from the DNA of UV-irradiated mammalian cells more efficiently than cyclobutane pyrimidine dimers," Mutation Research, vol. 143, no. 3, pp. 109–112, 1985.

115. W. H. Jeffrey, P. Aas, M. M. Lyons, R. B. Coffin, R. J. Pledger, and D. L. Mitchell, "Ambient solar radiation-induced photodamage in marine bacterioplankton," Photochemistry and Photobiology, vol. 64, no. 3, pp. 419–427, 1996.

116. H. C. Birnboim and J. J. Jevcak, "Fluorometric method for rapid detection of DNA strand breaks in human white blood cells produced by low doses of radiation," Cancer Research, vol. 41, no. 5, pp. 1889–1892, 1981.

117. T. Iwasa, S. Tokutomi, and F. Tokunaga, "Photoreactivation of Halobacterium halobium: action spectrum and role of pigmentation," Photochemistry and Photobiology, vol. 47, pp. 267–270, 1988.

118. A. Yasui, A. P. M. Eker, S. Yasuhira et al., "A new class of DNA photolyases present in various organisms including aplacental mammals," EMBO Journal, vol. 13, no. 24, pp. 6143–6151, 1994.

119. A. Kiener, I. Husain, A. Sancar, and C. Walsh, "Purification and properties of Methanobacterium thermoautotrophicum DNA photolyase," Journal of Biological Chemistry, vol. 264, no. 23, pp. 13880–13887, 1989.

120. M. Fujihashi, N. Numoto, Y. Kobayashi et al., "Crystal structure of archaeal photolyase from Sulfolobus tokodaii with two FAD molecules: implication of a novel light-harvesting cofactor," Journal of Molecular Biology, vol. 365, no. 4, pp. 903–910, 2007

121. K. Malhotra, S.-T. Kim, and A. Sancar, "Characterization of a medium wavelength type DNA photolyase: purification and properties of photolyase from Bacillus firmus," Biochemistry, vol. 33, no. 29, pp. 8712–8718, 1994

122. J. L. Johnson, S. Hamm-Alvarez, G. Payne, G. B. Sancar, K. V. Rajagopalan, and A. Sancar, "Identification of the second chromophore of Escherichia coli and yeast DNA photolyases as 5,10-methenyltetrahydrofolate," Proceedings of the National Academy of Sciences of the United States of America, vol. 85, no. 7, pp. 2046–2050, 1988

123. G. B. Sancar, F. W. Smith, M. C. Lorence, C. S. Rupert, and A. Sancar, "Sequences of the Escherichia coliphotolyase gene and protein," Journal of Biological Chemistry, vol. 259, no. 9, pp. 6033–6038, 1984.

124. A. P. M. Eker, P. Kooiman, J. K. C. Hessels, and A. Yasui, "DNA photoreactivating enzyme from the cyanobacterium Anacystis nidulans," Journal of Biological Chemistry, vol. 265, no. 14, pp. 8009–8015, 1990.

125. W.-O. Ng, R. Zentella, Y. Wang, J.-S. A. Taylor, and H. B. Pakrasi, "phrA, the major photoreactivating factor in the cyanobacterium Synechocystis sp. strain PCC 6803 codes for a cyclobutane-pyrimidine-dimer- specific DNA photolyase," Archives of Microbiology, vol. 173, no. 5-6, pp. 412–417, 2000

126. A. P. Eker, R. H. Dekker, and W. Berends, "Photoreactivating enzyme from Streptomyces griseus-IV. On the nature of the chromophoric cofactor in Streptomyces griseus photoreactivating enzyme,"Photochemistry and Photobiology, vol. 33, no. 1, pp. 65–72, 1981.

127. K. A. O›Connor, M. J. McBride, M. West et al., "Photolyase of Myxococcus xanthus, a Gram-negative eubacterium, is more

similar to photolyases found in archaea and «higher» eukaryotes than to photolyases of other eubacteria," Journal of Biological Chemistry, vol. 271, no. 11, pp. 6252–6259, 1996.

128. E. N. Worthington, I. H. Kavakli, G. Berrocal-Tito, B. E. Bondo, and A. Sancar, "Purification and characterization of three members of the photolyase/cryptochrome family blue-light photoreceptors from Vibrio cholerae," Journal of Biological Chemistry, vol. 278, no. 40, pp. 39143–39154, 2003

129. H. Yajima, H. Inoue, A. Oikawa, and A. Yasui, "Cloning and functional characterization of a eucaryotic DNA photolyase gene from Neurospora crassa," Nucleic Acids Research, vol. 19, no. 19, pp. 5359–5362, 1991.

130. A. P. M. Eker, H. Yajima, and A. Yasui, "DNA photolyase from the fungus Neurospora crassa. Purification, characterization and comparison with other photolyases," Photochemistry and Photobiology, vol. 60, no. 2, pp. 125–133, 1994.

131. T. Todo, H. Takemori, H. Ryo et al., "A new photoreactivating enzyme that specifically repairs ultraviolet light-induced (6–4) photoproducts," Nature, vol. 361, no. 6410, pp. 371–374, 1993

132. A. Sancar, "No «End of history» for photolyases," Science, vol. 272, no. 5258, pp. 48–49, 1996.

133. S. Yasuhira and A. Yasui, "Visible light-inducible photolyase gene from the goldfish Carassius auratus,"Journal of Biological Chemistry, vol. 267, no. 36, pp. 25644–25647, 1992.

134. T. S. Mikkelsen, M. J. Wakefield, B. Aken et al., "Genome of the marsupial Monodelphis domesticareveals innovation in non-coding sequences," Nature, vol. 447, no. 7141, pp. 167–177, 2007

135. T. Todo, S.-T. Kim, K. Hitomi et al., "Flavin adenine dinucleotide as a chromophore of the Xenopus (6–4)photolyase," Nucleic Acids Research, vol. 25, no. 4, pp. 764–768, 1997

136. Q. Pang and J. B. Hays, "UV-B-inducible and temperature-sensitive photoreactivation of cyclobutane pyrimidine dimers in Arabidopsis thaliana," Plant Physiology, vol. 95, no. 2, pp. 536–543, 1991.

137. J. L. Petersen, D. W. Lang, and G. D. Small, "Cloning and characterization of a class II DNA photolyase from

Chlamydomonas," Plant Molecular Biology, vol. 40, no. 6, pp. 1063–1071, 1999

138. S. Takahashi, N. Nakajima, H. Saji, and N. Kondo, "Diurnal changes of cucumber CPD photolyase gene (CsPHR) expression and its physiological role in growth under UV-B irradiation," Plant and Cell Physiology, vol. 43, pp. 342–349, 2002.
139. Y. Takeuchi, T. Inoue, K. Takemura, et al., "Induction and inhibition of cyclobutane pyrimidine dimer photolyase in etiolated cucumber (Cucumis sativus) cotyledons after ultraviolet irradiation depends on wavelength," Journal of Plant Research, vol. 120, pp. 365–374, 2007.
140. J. E. Trosko and V. H. Mansour, "Photoreactivation of ultraviolet light-induced pyrimidine dimers inGinkgo cells grown in vitro," Mutation Research, vol. 7, no. 1, pp. 120–121, 1969.
141. R. Taylor, A. Tobin, and C. Bray, "Nucleotide sequence of an Arabidopsis cDNA At-phrII encoding a protein with high homology to the class II CPD photolyases present in higher eukaryotes," Plant Physiology, vol. 112, pp. 862–863, 1996.
142. V. Srinivasan, W. M. Schnitzlein, and D. N. Tripathy, "Fowlpox virus encodes a novel DNA repair enzyme, CPD-photolyase, that restores infectivity of UV light-damaged virus," Journal of Virology, vol. 75, no. 4, pp. 1681–1688, 2001
143. C. L. Afonso, E. R. Tulman, Z. Lu, E. Oma, G. F. Kutish, and D. L. Rock, "The genome of Melanoplus sanguinipes entomopoxvirus," Journal ofVirology, vol. 73, no. 1, pp. 533–552, 1999.
144. T. Ohnishi, E. Mori, and A. Takahashi, "DNA double-strand breaks: their production, recognition, and repair in eukaryotes," Mutation Research, vol. 669, no. 1-2, pp. 8–12, 2009
145. D. Strumberg, A. A. Pilon, M. Smith, R. Hickey, L. Malkas, and Y. Pommier, "Conversion of topoisomerase I cleavage complexes on the leading strand of ribosomal DNA into 5›-phosphorylated DNA double-strand breaks by replication runoff," Molecular and Cellular Biology, vol. 20, no. 11, pp. 3977–3987, 2000
146. H. C. Box, J. B. Dawidzik, and E. E. Budzinski, "Free radical-induced double lesions in DNA," Free Radical Biology and Medicine, vol. 31, no. 7, pp. 856–868, 2001

147. J. P. Banáth and P. L. Olive, "Expression of phosphorylated histone H2AX as a surrogate of cell killing by drugs that create DNA double-strand breaks," Cancer Research, vol. 63, no. 15, pp. 4347–4350, 2003.
148. J. V. Harper, J. A. Anderson, and P. O'Neill, "Radiation induced DNA DSBs: contribution from stalled replication forks?" DNA Repair, vol. 9, pp. 907–913, 2010.
149. S. E. Freeman, A. D. Blackett, and D. C. Monteleone, "Quantitation of radiation-, chemical-, or enzyme-induced single strand breaks in nonradioactive DNA by alkaline gel electrophoresis: application to pyrimidine dimers," Analytical Biochemistry, vol. 158, no. 1, pp. 119–129, 1986.
150. D. L. Mitchell, J. Jen, and J. E. Cleaver, "Relative induction of cyclobutane dimers and cytosine photohydrates in DNA irradiated in vitro and in vivo with ultraviolet-C and ultraviolet-B light,"Photochemistry and Photobiology, vol. 54, no. 5, pp. 741–746, 1991.
151. B. Thyagarajan, K. E. Anderson, C. J. Lessard et al., "Alkaline unwinding flow cytometry assay to measure nucleotide excision repair," Mutagenesis, vol. 22, no. 2, pp. 147–153, 2007
152. P. Kara, K. Da□deviren, and M. Özsöz, "An electrochemical DNA biosensor for the detection of DNA damage caused by radioactive iodine and technetium," Turkish Journal of Chemistry, vol. 31, no. 3, pp. 243–249, 2007.
153. A. Hollaender and J. T. Curtis, "Effects of sublethal doses of monochromatic ultraviolet radiation on bacteria in liquid suspension," Proceedings of the Society for Experimental Biology and Medicine, vol. 33, pp. 61–62, 1935.
154. A. Kelner, "Effects of visible light on the recovery of Streptomyces griseus conidia from ultraviolet irradiation injury," Proceedings of the National Academy of Sciences of the Unites States of America, vol. 35, pp. 73–79, 1949.
155. R. Dulbecco, "Reactivation of ultra-violet-inactivated bacteriophage by visible light," Nature, vol. 163, no. 4155, pp. 949–950, 1949.
156. J. Essers, W. Vermeulen, and A. B. Houtsmuller, "DNA damage repair: anytime, anywhere?" Current Opinion in Cell Biology, vol. 18, no. 3, pp. 240–246, 2006

157. S.-T. Kim, P. F. Heelis, and A. Sancar, "Energy transfer (deazaflavin → FADH2) and electron transfer (FADH2 → T<>T) kinetics in Anacystis nidulans photolyase," Biochemistry, vol. 31, no. 45, pp. 11244–11248, 1992
158. L. O. Essen and T. Klar, "Light-driven DNA repair by photolyases," Cellular and Molecular Life Sciences, vol. 63, no. 11, pp. 1266–1277, 2006
159. A. B. Britt, "DNA damage and repair in plants," Annual Review of Plant Physiology and Plant Molecular Biology, vol. 47, no. 1, pp. 75–100, 1996.
160. T. Todo, "Functional diversity of the DNA photolyase/blue light receptor family," Mutation Research, vol. 434, no. 2, pp. 89–97, 1999
161. B. M. Sutherland, "Photoreactivating enzyme from human leukocytes," Nature, vol. 248, no. 5444, pp. 109–112, 1974.
162. B. M. Sutherland and P. V. Bennett, "Human white blood cells contain cyclobutyl pyrimidine dimer photolyase," Proceedings of the National Academy of Sciences of the United States of America, vol. 92, no. 21, pp. 9732–9736, 1995
163. H. Harm, "Damage and repair in mammalian cells after exposure to non-ionizing radiations. III. Ultraviolet and visible light irradiation of cells of placental mammals, including humans, and determination of photorepairable damage in vitro," Mutation Research, vol. 69, no. 1, pp. 167–176, 1980.
164. D. S. Hsu, X. Zhao, S. Zhao et al., "Putative human blue-light photoreceptors hCRY1 and hCRY2 are flavoproteins," Biochemistry, vol. 35, no. 44, pp. 13871–13877, 1996
165. T. Ueda, A. Kato, S. Kuramitsu, H. Terasawa, and I. Shimada, "Identification and characterization of a second chromophore of DNA photolyase from Thermus thermophilus HB27," Journal of Biological Chemistry, vol. 280, no. 43, pp. 36237–36243, 2005
166. C. Saxena, A. Sancar, and D. Zhong, "Femtosecond dynamics of DNA photolyase: energy transfer of antenna initiation and electron transfer of cofactor reduction," Journal of Physical Chemistry B, vol. 108, no. 46, pp. 18026–18033, 2004
167. Y.-T. Kao, C. Saxena, L. Wang, A. Sancar, and D. Zhong, "Direct observation of thymine dimer repair in DNA by photolyase,"

Proceedings of the National Academy of Sciences of the United States of America, vol. 102, no. 45, pp. 16128–16132, 2005

168. A. E. Stapleton and V. Walbot, "Flavonoids can protect maize DNA from the induction of ultraviolet radiation damage," Plant Physiology, vol. 105, no. 3, pp. 881–889, 1994.
169. R. Schmitz-Hoerner and G. Weissenbock, "Contribution of phenolic compounds to the UV-B screening capacity of developing barley primary leaves in relation to DNA damage and repair under elevated UV-B levels," Phytochemistry, vol. 64, pp. 243–255, 2003.
170. G. Kaiser, O. Kleiner, C. Beisswenger, and A. Batschauer, "Increased DNA repair in Arabidopsis plants overexpressing CPD photolyase," Planta, vol. 230, no. 3, pp. 505–515, 2009
171. N. Tuteja, M. B. Singh, M. K. Misra, P. L. Bhalla, and R. Tuteja, "Molecular mechanisms of DNA damage and repair: progress in plants," Critical Reviews in Biochemistry and Molecular Biology, vol. 36, no. 4, pp. 337–397, 2001.
172. S. Kimura, Y. Tahira, T. Ishibashi, et al., "DNA repair in higher plants; photoreactivation is the major DNA repair pathway in non-proliferating cells while excision repair (nucleotide excision repair and base excision repair) is active in proliferating cells," Nucleic Acids Research, vol. 32, pp. 2760–2767, 2004.
173. J. J. Chen, D. L. Mitchell, and A. B. Britt, "A light-dependent pathway for the elimination of UV-induced pyrimidine (6–4) pyrimidinone photoproducts in Arabidopsis," Plant Cell, vol. 6, no. 9, pp. 1311–1317, 1994
174. H.-S. Kang, J. Hidema, and T. Kumagai, "Effects of light environment during culture on UV-induced cyclobutyl pyrimidine dimers and their photorepair in rice (Oryza sativa L.)," Photochemistry and Photobiology, vol. 68, no. 1, pp. 71–77, 1998.
175. A. Mees, T. Klar, P. Gnau et al., "Crystal structure of a photolyase bound to a CPD-like DNA lesion after in situ repair," Science, vol. 306, no. 5702, pp. 1789–1793, 2004
176. T. Torizawa, T. Ueda, S. Kuramitsu et al., "Investigation of the cyclobutane pyrimidine dimer (CPD) photolyase DNA recognition mechanism by NMR analyses," Journal of Biological Chemistry, vol. 279, no. 31, pp. 32950–32956, 2004

177. P. F. Heelis, R. F. Hartman, and S. D. Rose, "Photoenzymic repair of UV-damaged DNA—a chemist's perspective," Chemical Society Reviews, vol. 24, no. 4, pp. 289–297, 1995.
178. A. W. MacFarlane and R. J. Stanley, "Cis-Syn thymidine dimer repair by DNA photolyase in real time,"Biochemistry, vol. 42, no. 28, pp. 8558–8568, 2003
179. I. H. Kavakli and A. Sancar, "Analysis of the role of intraprotein electron transfer in photoreactivation by DNA photolyase in vivo," Biochemistry, vol. 43, no. 48, pp. 15103–15110, 2004
180. E. Schleicher, B. Heßling, V. Illarionova et al., "Light-induced reactions of Escherichia coli DNA photolyase monitored by Fourier transform infrared spectroscopy," FEBS Journal, vol. 272, no. 8, pp. 1855–1866, 2005
181. G. B. Sancar and F. W. Smith, "Interactions between yeast photolyase and nucleotide excision repair proteins in Saccharomyces cerevisiae and Escherichia coli," Molecular and Cellular Biology, vol. 9, no. 11, pp. 4767–4776, 1989.
182. E. Seeberg, L. Eide, and M. Bjørås, "The base excision repair pathway," Trends in Biochemical Sciences, vol. 20, no. 10, pp. 391–397, 1995
183. J. H. Houtgraaf, J. Versmissen, and W. J. van der Giessen, "A concise review of DNA damage checkpoints and repair in mammalian cells," Cardiovascular Revascularization Medicine, vol. 7, no. 3, pp. 165–172, 2006
184. K. H. Almeida and R. W. Sobol, "A unified view of base excision repair: lesion-dependent protein complexes regulated by post-translational modification," DNA Repair, vol. 6, no. 6, pp. 695–711, 2007
185. K. Hitomi, S. Iwai, and J. A. Tainer, "The intricate structural chemistry of base excision repair machinery: implications for DNA damage recognition, removal, and repair," DNA Repair, vol. 6, no. 4, pp. 410–428, 2007
186. A. B. Robertson, A. Klungland, T. Rognes, and I. Leiros, "Base excision repair: the long and short of it,"Cellular and Molecular Life Sciences, vol. 66, no. 6, pp. 981–993, 2009
187. D. E. Barnes and T. Lindahl, "Repair and genetic consequences of endogenous DNA base damage in mammalian cells," Annual Review of Genetics, vol. 38, pp. 445–476, 2004

188. J. L. Parsons, P. S. Tait, D. Finch, I. I. Dianova, S. L. Allinson, and G. L. Dianov, "CHIP-mediated degradation and DNA damage-dependent stabilization regulate base excision repair proteins,"Molecular Cell, vol. 29, no. 4, pp. 477–487, 2008
189. K. W. Caldecott, "XRCC1 and DNA strand break repair," DNA Repair, vol. 2, no. 9, pp. 955–969, 2003
190. L. Lan, S. Nakajima, Y. Oohata et al., "In situ analysis of repair processes for oxidative DNA damage in mammalian cells," Proceedings of the National Academy of Sciences of the United States of America, vol. 101, no. 38, pp. 13738–13743, 2004
191. K. Sakumi and M. Sekiguchi, "Structures and functions of DNA glycosylases," Mutation Research, vol. 236, no. 2-3, pp. 161–172, 1990.
192. Y. Matsumoto and K. Kim, "Excision of deoxyribose phosphate residues by DNA polymerase β during DNA repair," Science, vol. 269, no. 5224, pp. 699–702, 1995.
193. B. Demple and L. Harrison, "Repair of oxidative damage to DNA: enzymology and biology," Annual Review of Biochemistry, vol. 63, pp. 915–948, 1994.
194. L. Gros, A. A. Ishchenko, H. Ide, R. H. Elder, and M. K. Saparbaev, "The major human AP endonuclease (Ape1) is involved in the nucleotide incision repair pathway," Nucleic Acids Research, vol. 32, no. 1, pp. 73–81, 2004. · View at Google Scholar
195. P. Fortini and E. Dogliotti, "Base damage and single-strand break repair: mechanisms and functional significance of short- and long-patch repair subpathways," DNA Repair, vol. 6, no. 4, pp. 398–409, 2007
196. U. Sattler, P. Frit, B. Salles, and P. Calsou, "Long-patch DNA repair synthesis during base excision repair in mammalian cells," EMBO Reports, vol. 4, no. 4, pp. 363–367, 2003
197. W. D. Henner, S. M. Grunberg, and W. A. Haseltine, "Enzyme action at 3› termini of ionizing radiation-induced DNA strand breaks," Journal of Biological Chemistry, vol. 258, no. 24, pp. 15198–151205, 1983.
198. W. D. Henner, L. O. Rodriguez, S. M. Hecht, and W. A. Haseltine, "gamma Ray induced deoxyribonucleic acid strand breaks. 3′ Glycolate termini," Journal of Biological Chemistry, vol. 258, no. 2, pp. 711–713, 1983.

199. A. Jilani, D. Ramotar, C. Slack et al., "Molecular cloning of the human gene, PNKP, encoding a polynucleotide kinase 3'-phosphatase and evidence for its role in repair of DNA strand breaks caused by oxidative damage," Journal of Biological Chemistry, vol. 274, no. 34, pp. 24176–24186, 1999

200. R. Gary, K. Kim, H. L. Cornelius, M. S. Park, and Y. Matsumoto, "Proliferating cell nuclear antigen facilitates excision in long-patch base excision repair," Journal of Biological Chemistry, vol. 274, no. 7, pp. 4354–4363, 1999

201. P. Fortini, B. Pascucci, E. Parlanti, R. W. Sobol, S. H. Wilson, and E. Dogliotti, "Different DNA polymerases are involved in the short- and long-patch base excision repair in mammalian cells,"Biochemistry, vol. 37, no. 11, pp. 3575–3580, 1998

202. A. B. Britt, "Molecular genetics of DNA repair in higher plants," Trends in Plant Science, vol. 4, no. 1, pp. 20–25, 1999

203. J. B. Hays, "Arabidopsis thaliana, a versatile model system for study of eukaryotic genome-maintenance functions," DNA Repair, vol. 1, no. 8, pp. 579–600, 2002

204. E. Hefner, S. B. Preuss, and A. B. Britt, "Arabidopsis mutants sensitive to gamma radiation include the homologue of the human repair gene ERCC1," Journal of Experimental Botany, vol. 54, no. 383, pp. 669–680, 2003

205. A. G. McLennan, "The repair of ultraviolet light-induced DNA damage in plant cells," Mutation Research, vol. 181, no. 1, pp. 1–7, 1987.

206. E. J. Vonarx, H. L. Mitchell, R. Karthikeyan, I. Chatterjee, and B. A. Kunz, "DNA repair in higher plants," Mutation Research, vol. 400, no. 1-2, pp. 187–200, 1998

207. E. Dandoy, R. Schyns, R. Deltour, and W. G. Verly, "Appearance and repair of apurinic/apyrimidinic sites in DNA during early germination of Zea mays," Mutation Research, vol. 181, no. 1, pp. 57–60, 1987.

208. M. Talpaert-Borlè and M. Liuzzi, "Base-excision repair in carrot cells," European Journal of Biochemistry, vol. 124, no. 3, pp. 435–440, 1982.

209. R. T. Hess, U. Schwitter, M. Petretta, B. Giese, and H. Naegeli, "Bipartite substrate discrimination by human nucleotide excision

repair," Proceedings of the National Academy of Sciences of the United States of America, vol. 94, no. 13, pp. 6664–6669, 1997

210. R. M. A. Costa, V. Chiganças, R. D. S. Galhardo, H. Carvalho, and C. F. M. Menck, "The eukaryotic nucleotide excision repair pathway," Biochimie, vol. 85, no. 11, pp. 1083–1099, 2003

211. A. Van Hoffen, J. Venema, R. Meschini, A. A. Van Zeeland, and L. H. F. Mullenders, "Tvanscription-coupled repair removes both cyclobutane pyrimidine dimers and 6–4 photoproducts with equal efficiency and in a sequential way from transcribed DNA in xeroderma pigmentosum group C fibroblasts," EMBO Journal, vol. 14, no. 2, pp. 360–367, 1995.

212. R. P. Boyce and P. Howard-Flanders, "Release of ultraviolet light-induced thymine dimers from DNA inE. coli K-12," Proceedings of the National Academy of Sciences of the United States of America, vol. 51, pp. 293–300, 1964.

213. D. E. Pettijohn and P. C. Hanawalt, "Evidence for repair-replication of ultraviolet damaged DNA in bacteria," Journal of Molecular Biology, vol. 9, pp. 395–410, 1964.

214. J.-J. Lin and A. Sancar, "Reconstitution of nucleotide excision nuclease with UvrA and UvrB proteins from Escherichia coli and UvrC protein from Bacillus subtilis," Journal of Biological Chemistry, vol. 265, no. 34, pp. 21337–21341, 1990.

215. D. K. Orren, P. Selby, J. E. Hearst, and A. Sancar, "Post-incision steps of nucleotide excision repair inEscherichia coli. Disassembly of the UvrBC-DNA complex by helicase II and DNA polymerase I," The Journal of Biological Chemistry, vol. 267, pp. 780–788, 1992.

216. T. Nouspikel, "Nucleotide excision repair: variations on versatility," Cellular and Molecular Life Sciences, vol. 66, no. 6, pp. 994–1009, 2009.

217. S. W. Matson, D. W. Bean, and J. W. George, "DNA helicases: enzymes with essential roles in all aspects of DNA metabolism," BioEssays, vol. 16, no. 1, pp. 13–22, 1994.

218. K. A. Henning, L. Li, N. Iyer et al., "The Cockayne syndrome group a gene encodes a WD repeat protein that interacts with CSB protein and a subunit of RNA polymerase II TFIIH," Cell, vol. 82, no. 4, pp. 555–564, 1995.

219. C. P. Selby and A. Sancar, "Human transcription-repair coupling factor CSB/ERCC6 is a DNA-stimulated ATPase but is not a helicase and does not disrupt the ternary transcription complex of stalled RNA polymerase II," Journal of Biological Chemistry, vol. 272, no. 3, pp. 1885–1890, 1997
220. P. C. Hanawalt, B. A. Donahue, and K. S. Sweder, "Repair and transcription: collision or collusion?"Current Biology, vol. 4, no. 6, pp. 518–522, 1994
221. S. Tornaletti and P. C. Hanawalt, "Effect of DNA lesions on transcription elongation," Biochimie, vol. 81, no. 1-2, pp. 139–146, 1999
222. J. N. Ratner, B. Balasubramanian, J. Corden, S. L. Warren, and D. B. Bregman, "Ultraviolet radiation-induced ubiquitination and proteasomal degradation of the large subunit of RNA polymerase II: implications for transcription-coupled DNA repair," Journal of Biological Chemistry, vol. 273, no. 9, pp. 5184–5189, 1998
223. A. Sarasin and A. Stary, "New insights for understanding the transcription-coupled repair pathway,"DNA Repair, vol. 6, no. 2, pp. 265–269, 2007
224. M. Fousteri, W. Vermeulen, A. A. van Zeeland, and L. H. F. Mullenders, "Cockayne syndrome A and B proteins differentially regulate recruitment of chromatin remodeling and repair factors to stalled RNA polymerase II in vivo," Molecular Cell, vol. 23, no. 4, pp. 471–482, 2006
225. L. Proietti-De-Santis, P. Drané, and J.-M. Egly, "Cockayne syndrome B protein regulates the transcriptional program after UV irradiation," EMBO Journal, vol. 25, no. 9, pp. 1915–1923, 2006
226. K. Sugasawa, J. M. Y. Ng, C. Masutani et al., "Xeroderma pigmentosum group C protein complex is the initiator of global genome nucleotide excision repair," Molecular Cell, vol. 2, no. 2, pp. 223–232, 1998.
227. W. L. De Laat, N. G. J. Jaspers, and J. H. J. Hoeijmakers, "Molecular mechanism of nucleotide excision repair," Genes and Development, vol. 13, no. 7, pp. 768–785, 1999.
228. L. C. J. Gillet and O. D. Schärer, "Molecular mechanisms of mammalian global genome nucleotide excision repair," Chemical Reviews, vol. 106, no. 2, pp. 253–276, 2006

229. C. Petit and A. Sancar, "Nucleotide excision repair: from E. coli to man," Biochimie, vol. 81, no. 1-2, pp. 15–25, 1999
230. C. Masutani, K. Sugasawa, J. Yanagisawa et al., "Purification and cloning of a nucleotide excision repair complex involving the xeroderma pigmentosum group C protein and a human homologue of yeast RAD23," EMBO Journal, vol. 13, no. 8, pp. 1831–1843, 1994.
231. M. Araki, C. Masutani, M. Takemura et al., "Centrosome protein centrin 2/caltractin 1 is part of the xeroderma pigmentosum group C complex that initiates global genome nucleotide excision repair,"Journal of Biological Chemistry, vol. 276, no. 22, pp. 18665–18672, 2001
232. L. Schaeffer, R. Roy, S. Humbert, et al., "DNA repair helicase: a component of BTF2 (TFIIH) basic transcription factor," Science, vol. 260, pp. 58–63, 1993.
233. F. Coin, V. Oksenych, and J.-M. Egly, "Distinct roles for the XPB/p52 and XPD/p44 subcomplexes of TFIIH in damaged DNA opening during nucleotide excision repair," Molecular Cell, vol. 26, no. 2, pp. 245–256, 2007
234. L. Schaeffer, V. Moncollin, R. Roy et al., "The ERCC2/DNA repair protein is associated with the class II BTF2/TFIIH transcription factor," EMBO Journal, vol. 13, no. 10, pp. 2388–2392, 1994.
235. J. Chen, S. Larochelle, X. Li, and B. Suter, "Xpd/Ercc2 regulates CAK activity and mitotic progression,"Nature, vol. 424, no. 6945, pp. 228–232, 2003
236. P. Schultz, S. Fribourg, A. Poterszman, V. Mallouh, D. Moras, and J. M. Egly, "Molecular structure of human TFIIH," Cell, vol. 102, no. 5, pp. 599–607, 2000.
237. J. A. Ranish, S. Hahn, Y. Lu et al., "Identification of TFB5, a new component of general transcription and DNA repair factor IIH," Nature Genetics, vol. 36, no. 7, pp. 707–713, 2004
238. C. J. Jones and R. D. Wood, "Preferential binding of the xeroderma pigmentosum group A complementing protein to damaged DNA," Biochemistry, vol. 32, no. 45, pp. 12096–12104, 1993
239. T. Matsuda, M. Saijo, I. Kuraoka et al., "DNA repair protein XPA binds replication protein A (RPA),"Journal of Biological Chemistry, vol. 270, no. 8, pp. 4152–4157, 1995

240. E. Stigger, R. Drissi, and S.-H. Lee, "Functional analysis of human replication protein A in nucleotide excision repair," Journal of Biological Chemistry, vol. 273, no. 15, pp. 9337–9343, 1998

241. Z. He, L. A. Henricksen, M. S. Wold, and C. J. Ingles, "RPA involvement in the damage-recognition and incision steps of nucleotide excision repair," Nature, vol. 374, no. 6522, pp. 566–569, 1995.

242. J. L. Burns, S. N. Guzder, P. Sung, S. Prakash, and L. Prakash, "An affinity of human replication protein A for ultraviolet-damaged DNA: implications for damage recognition in nucleotide excision repair,"Journal of Biological Chemistry, vol. 271, no. 20, pp. 11607–11610, 1996.

243. E. Evans, J. G. Moggs, J. R. Hwang, J.-M. Egly, and R. D. Wood, "Mechanism of open complex and dual incision formation by human nucleotide excision repair factors," EMBO Journal, vol. 16, no. 21, pp. 6559–6573, 1997

244. A. H. Sarker, S. E. Tsutakawa, S. Kostek et al., "Recognition of RNA polymerase II and transcription bubbles by XPG, CSB, and TFIIH: insights for transcription-coupled repair and Cockayne syndrome,"Molecular Cell, vol. 20, no. 2, pp. 187–198, 2005

245. K. G. Cloud, B. Shen, G. F. Strniste, and M. S. Park, "XPG protein has a structure-specific endonuclease activity," Mutation Research, vol. 347, no. 2, pp. 55–60, 1995.

246. W. L. De Laat, E. Appeldoorn, N. G. J. Jaspers, and J. H. J. Hoeijmakers, "DNA structural elements required for ERCC1-XPF endonuclease activity," Journal of Biological Chemistry, vol. 273, no. 14, pp. 7835–7842, 1998

247. J. H. Enzlin and O. D. Schärer, "The active site of the DNA repair endonuclease XPF-ERCC1 forms a highly conserved nuclease motif," EMBO Journal, vol. 21, no. 8, pp. 2045–2053, 2002

248. R. D. Wood, "Nucleotide excision repair in mammalian cells," Journal of Biological Chemistry, vol. 272, no. 38, pp. 23465–23468, 1997

249. D. Mu, D. S. Hsu, and A. Sancar, "Reaction mechanism of human DNA repair excision nuclease,"Journal of Biological Chemistry, vol. 271, no. 14, pp. 8285–8294, 1996

250. K. Sugasawa, Y. Okuda, M. Saijo et al., "UV-induced ubiquitylation of XPC protein mediated by UV-DDB-ubiquitin ligase complex," Cell, vol. 121, no. 3, pp. 387–400, 2005
251. J. Tang and G. Chu, "Xeroderma pigmentosum complementation group E and UV-damaged DNA-binding protein," DNA Repair, vol. 1, no. 8, pp. 601–616, 2002
252. O. Maillard, S. Solyom, and H. Naegeli, "An aromatic sensor with aversion to damaged strands confers versatility to DNA repair," PLoS Biology, vol. 5, no. 4, article no. e79, 2007
253. K. Sugasawa, "UV-induced ubiquitylation of XPC complex, the UV-DDB-ubiquitin ligase complex, and DNA repair," Journal of Molecular Histology, vol. 37, no. 5-7, pp. 189–202, 2006
254. D. Hoogstraten, S. Bergink, V. H. M. Verbiest et al., "Versatile DNA damage detection by the global genome nucleotide excision repair protein XPC," Journal of Cell Science, vol. 121, no. 17, pp. 2850–2859, 2008
255. P. J. Van der Spek, A. Eker, S. Rademakers et al., "XPC and human homologs of RAD23: intracellular localization and relationship to other nucleotide excision repair complexes," Nucleic Acids Research, vol. 24, no. 13, pp. 2551–2559, 1996
256. K. Sugasawa, C. Masutani, A. Uchida et al., "HHR23B, a human Rad23 homolog, stimulates XPC protein in nucleotide excision repair in vitro," Molecular and Cellular Biology, vol. 16, no. 9, pp. 4852–4861, 1996.
257. C. Masutani, M. Araki, K. Sugasawa et al., "Identification and characterization of XPC-binding domain of hHR23B," Molecular and Cellular Biology, vol. 17, no. 12, pp. 6915–6923, 1997.
258. K. Sugasawa, J. M. Y. Ng, C. Masutani et al., "Two human homologs of Rad23 are functionally interchangeable in complex formation and stimulation of XPC repair activity," Molecular and Cellular Biology, vol. 17, no. 12, pp. 6924–6931, 1997.
259. J. T. Reardon, D. Mu, and A. Sancar, "Overproduction, purification, and characterization of the XPC subunit of the human DNA repair excision nuclease," Journal of Biological Chemistry, vol. 271, no. 32, pp. 19451–19456, 1996
260. K. Sugasawa, T. Okamoto, Y. Shimizu, C. Masutani, S. Iwai, and F. Hanaoka, "A multistep damage recognition mechanism for global

genomic nucleotide excision repair," Genes and Development, vol. 15, no. 5, pp. 507–521, 2001

261. A. Aboussekhra, M. Biggerstaff, M. K. K. Shivji et al., "Mammalian DNA nucleotide excision repair reconstituted with purified protein components," Cell, vol. 80, no. 6, pp. 859–868, 1995.
262. F. Tirode, D. Busso, F. Coin, and J.-M. Egly, "Reconstitution of the transcription factor TFIIH: assignment of functions for the three enzymatic subunits, XPB, XPD, and cdk7," Molecular Cell, vol. 3, no. 1, pp. 87–95, 1999
263. D. Mu, C.-H. Park, T. Matsunaga, D. S. Hsu, J. T. Reardon, and A. Sancar, "Reconstitution of human DNA repair excision nuclease in a highly defined system," Journal of Biological Chemistry, vol. 270, no. 6, pp. 2415–2418, 1995
264. M. R. Lieber, "The FEN-1 family of structure-specific nucleases in eukaryotic DNA replication, recombination and repair," BioEssays, vol. 19, no. 3, pp. 233–240, 1997.
265. A. Constantinou, D. Gunz, E. Evans et al., "Conserved residues of human XPG protein important for nuclease activity and function in nucleotide excision repair," Journal of Biological Chemistry, vol. 274, no. 9, pp. 5637–5648, 1999
266. R. Guo, J. Chen, F. Zhu, et al., "E2F1 Localizes to sites of UV-induced DNA damage to enhance nucleotide excision repair," The Journal of Biological Chemistry, vol. 285, pp. 19308–19315, 2010
267. S. Nocentini, "Rejoining kinetics of DNA single- and double-strand breaks in normal and DNA ligase-deficient cells after exposure to ultraviolet C and gamma radiation: an evaluation of ligating activities involved in different DNA repair processes," Radiation Research, vol. 151, no. 4, pp. 423–432, 1999.
268. J. Moser, H. Kool, I. Giakzidis, K. Caldecott, L. H.F. Mullenders, and M. I. Fousteri, "Sealing of chromosomal DNA nicks during nucleotide excision repair requires XRCC1 and DNA ligase IIIα in a cell-cycle-specific manner," Molecular Cell, vol. 27, no. 2, pp. 311–323, 2007
269. S. Tornaletti, "Transcription-coupled DNA repair: directing your effort where it›s most needed," Cellular and Molecular Life Sciences, vol. 66, no. 6, pp. 1010–1020, 2009

270. G. R. Harlow, M. E. Jenkins, T. S. Pittalwala, and D. W. Mount, "Isolation of uvh1, an Arabidopsis mutant hypersensitive to ultraviolet light and ionizing radiation," Plant Cell, vol. 6, no. 2, pp. 227–235, 1994

271. A. L. Fidantsef, D. L. Mitchell, and A. B. Britt, "The Arabidopsis UVH1 gene is a homolog of the yeast repair endonuclease RAD1," Plant Physiology, vol. 124, no. 2, pp. 579–586, 2000.

272. Z. Liu, G. S. Hossain, M. A. Islas-Osuna, D. L. Mitchell, and D. W. Mount, "Repair of UV damage in plants by nucleotide excision repair: Arabidopsis UVH1 DNA repair gene is a homolog of Saccharomyces cerevisiae Rad1," Plant Journal, vol. 21, no. 6, pp. 519–528, 2000

273. Z. Liu, J. D. Hall, and D. W. Mount, "Arabidopsis UVH3 gene is a homolog of the Saccharomyces cerevisiae RAD2 and human XPG DNA repair genes," Plant Journal, vol. 26, no. 3, pp. 329–338, 2001

274. G. C. Cannon, L. A. Hedrick, and S. Heinhorst, "Repair mechanisms of UV-induced DNA damage in soybean chloroplasts," Plant Molecular Biology, vol. 29, no. 6, pp. 1267–1277, 1995.

275. J. Hidema, T. Kumagai, J. C. Sutherland, and B. M. Sutherland, "Ultraviolet B-sensitive rice cultivar deficient in cyclobutyl pyrimidine dimer repair," Plant Physiology, vol. 113, no. 1, pp. 39–44, 1997.

276. M. E. Jenkins, G. R. Harlow, Z. Liu, M. A. Shotwell, J. Ma, and D. W. Mount, "Radiation-sensitive mutants of Arabidopsis thaliana," Genetics, vol. 140, no. 2, pp. 725–732, 1995.

277. C.-Z. Jiang, C.-N. Yen, K. Cronin, D. Mitchell, and A. B. Britt, "UV- and gamma-radiation sensitive mutants of Arabidopsis thaliana," Genetics, vol. 147, no. 3, pp. 1401–1409, 1997.

278. S. Kimura and K. Sakaguchi, "DNA repair in plants," Chemical Reviews, vol. 106, no. 2, pp. 753–766, 2006

279. H. Xu, I. Swoboda, P. L. Bhalla et al., "Plant homologue of human excision repair gene ERCC1 points to conservation of DNA repair mechanisms," Plant Journal, vol. 13, no. 6, pp. 823–829, 1998

280. E. C. Friedberg, "DNA damage and repair," Nature, vol. 421, no. 6921, pp. 436–440, 2003

281. J. Wu, L.-Y. Lu, and X. Yu, "The role of BRCA1 in DNA damage response," Protein Cell, vol. 1, pp. 117–123, 2010.

282. P. Sung, K. M. Trujillo, and S. Van Komen, "Recombination factors of Saccharomyces cerevisiae,"Mutation Research, vol. 451, no. 1-2, pp. 257–275, 2000

283. G. A. Cromie, J. C. Connelly, and D. R. F. Leach, "Recombination at double-strand breaks and DNA ends: conserved mechanisms from phage to humans," Molecular Cell, vol. 8, no. 6, pp. 1163–1174, 2001

284. J. Thacker, "The RAD51 gene family, genetic instability and cancer," Cancer Letters, vol. 219, no. 2, pp. 125–135, 2005

285. K. McEntee, G. M. Weinstock, and I. R. Lehman, "recA protein-catalyzed strand assimilation: stimulation by Escherichia coli single-stranded DNA-binding protein," Proceedings of the National Academy of Sciences of the United States of America, vol. 77, no. 2, pp. 857–861, 1980.

286. D. K. Bishop, D. Park, L. Xu, and N. Kleckner, "DMC1: a meiosis-specific yeast homolog of E. coli recA required for recombination, synaptonemal complex formation, and cell cycle progression," Cell, vol. 69, no. 3, pp. 439–456, 1992

287. S. T. Lovett, "Sequence of the RAD5 gene of Saccharomyces cerevisiae: similarity of RAD55 to prokaryotic RecA and other RecA-like proteins," Gene, vol. 142, no. 1, pp. 103–106, 1994

288. W. Li and H. Ma, "Double-stranded DNA breaks and gene functions in recombination and meiosis,"Cell Research, vol. 16, no. 5, pp. 402–412, 2006

289. M. Kawabata, T. Kawabata, and M. Nishibori, "Role of recA/RAD51 family proteins in mammals," Acta Medica Okayama, vol. 59, no. 1, pp. 1–9, 2005.

290. Y. Aylon and M. Kupiec, "New insights into the mechanism of homologous recombination in yeast,"Mutation Research, vol. 566, no. 3, pp. 231–248, 2004.

291. M. S. Park, "Expression of human RAD52 confers resistance to ionizing radiation in mammalian cells,"Journal of Biological Chemistry, vol. 270, no. 26, pp. 15467–15470, 1995.

292. Y. Aylon and M. Kupiec, "The checkpoint protein Rad24 of Saccharomyces cerevisiae is involved in processing double-strand

break ends and in recombination partner choice," Molecular and Cellular Biology, vol. 23, no. 18, pp. 6585–6596, 2003

293. S. Palmer, E. Schildkraut, R. Lazarin, J. Nguyen, and J. A. Nickoloff, "Gene conversion tracts inSaccharomyces cerevisiae can be extremely short and highly directional," Nucleic Acids Research, vol. 31, no. 4, pp. 1164–1173, 2003

294. O. Inbar and M. Kupiec, "Homology search and choice of homologous partner during mitotic recombination," Molecular and Cellular Biology, vol. 19, no. 6, pp. 4134–4142, 1999.

295. Y. Aylon, B. Liefshitz, G. Bitan-Banin, and M. Kupiec, "Molecular dissection of mitotic recombination in the yeast Saccharomyces cerevisiae," Molecular and Cellular Biology, vol. 23, no. 4, pp. 1403–1417, 2003

296. H. Puchta, P. Swoboda, and B. Hohn, "Induction of intrachromosomal homologous recombination in whole plants," Plant Journal, vol. 7, no. 2, pp. 203–210, 1995.

297. P. A. Jeggo, "Studies on mammalian mutants defective in rejoining double-strand breaks in DNA,"Mutation Research, vol. 239, no. 1, pp. 1–16, 1990.

298. G. Ira, A. Pellicioll, A. Balijja et al., "DNA end resection, homologous recombination and DNA damage checkpoint activation require CDK1," Nature, vol. 431, no. 7011, pp. 1011–1017, 2004

299. P. Sung, "Yeast Rad55 and Rad57 proteins form a heterodimer that functions with replication protein A to promote DNA strand exchange by Rad51 recombinase," Genes and Development, vol. 11, no. 9, pp. 1111–1121, 1997.

300. B. Wolner, S. Van Komen, P. Sung, and C. L. Peterson, "Recruitment of the recombinational repair machinery to a DNA double-strand break in yeast," Molecular Cell, vol. 12, no. 1, pp. 221–232, 2003

301. E. Sonoda, G. Y. Zhao, M. Kohzaki et al., "Collaborative roles of γH2AX and the Rad51 paralog Xrcc3 in homologous recombinational repair," DNA Repair, vol. 6, no. 3, pp. 280–292, 2007

302. F. Pâques and J. E. Haber, "Multiple pathways of recombination induced by double-strand breaks in Saccharomyces cerevisiae,"

Microbiology and Molecular Biology Reviews, vol. 63, no. 2, pp. 349–404, 1999.

303. S. E. Critchlow and S. P. Jackson, "DNA end-joining: from yeast to man," Trends in Biochemical Sciences, vol. 23, no. 10, pp. 394–398, 1998

304. M. L. Hefferin and A. E. Tomkinson, "Mechanism of DNA double-strand break repair by non-homologous end joining," DNA Repair, vol. 4, no. 6, pp. 639–648, 2005

305. S. A. McElhinny, J. M. Havener, M. Garcia-Diaz et al., "A gradient of template dependence defines distinct biological roles for family X polymerases in nonhomologous end joining," Molecular Cell, vol. 19, no. 3, pp. 357–366, 2005

306. J.-P. Capp, F. Boudsocq, P. Bertrand et al., "The DNA polymerase λ is required for the repair of non-compatible DNA double strand breaks by NHEJ in mammalian cells," Nucleic Acids Research, vol. 34, no. 10, pp. 2998–3007, 2006

307. S. Thode, A. Schäfer, P. Pfeiffer, and W. Vielmetter, "A novel pathway of DNA end-to-end joining,"Cell, vol. 60, no. 6, pp. 921–928, 1990

308. R. B. Cary, S. R. Peterson, J. Wang, D. G. Bear, E. M. Bradbury, and D. J. Chen, "DNA looping by Ku and the DNA-dependent protein kinase," Proceedings of the National Academy of Sciences of the United States of America, vol. 94, no. 9, pp. 4267–4272, 1997

309. D. A. Ramsden and M. Geliert, "Ku protein stimulates DNA end joining by mammalian DNA ligases: a direct role for Ku in repair of DNA double-strand breaks," EMBO Journal, vol. 17, no. 2, pp. 609–614, 1998

310. P. Baumann and S. C. West, "DNA end-joining catalyzed by human cell-free extracts," Proceedings of the National Academy of Sciences of the United States of America, vol. 95, no. 24, pp. 14066–14070, 1998

311. H. Yajima, K.-J. Lee, S. Zhang, J. Kobayashi, and B. P. C. Chen, "DNA double-strand break formation upon UV-induced replication stress activates ATM and DNA-PKcs kinases," Journal of Molecular Biology, vol. 385, no. 3, pp. 800–810, 2009

312. C. Muller, P. Calsou, P. Frit, C. Cayrol, T. Carter, and B. Salles, "UV sensitivity and impaired nucleotide excision repair in DNA-

dependent protein kinase mutant cells," Nucleic Acids Research, vol. 26, no. 6, pp. 1382–1389, 1998

313. P. Hentges, P. Ahnesorg, R. S. Pitcher et al., "Evolutionary and functional conservation of the DNA non-homologous end-joining protein, XLF/Cernunnos," Journal of Biological Chemistry, vol. 281, no. 49, pp. 37517–37526, 2006
314. P. Ahnesorg, P. Smith, and S. P. Jackson, "XLF interacts with the XRCC4-DNA ligase IV complex to promote DNA nonhomologous end-joining," Cell, vol. 124, no. 2, pp. 301–313, 2006
315. D. Buck, L. Malivert, R. De Chasseval et al., "Cernunnos, a novel nonhomologous end-joining factor, is mutated in human immunodeficiency with microcephaly," Cell, vol. 124, no. 2, pp. 287–299, 2006
316. E. Riballo, M. Kühne, N. Rief et al., "A pathway of double-strand break rejoining dependent upon ATM, Artemis, and proteins locating to γ-H2AX foci," Molecular Cell, vol. 16, no. 5, pp. 715–724, 2004
317. J. Lloyd, J. R. Chapman, J. A. Clapperton et al., "A supramodular FHA/BRCT-repeat architecture mediates Nbs1 adaptor function in response to DNA damage," Cell, vol. 139, no. 1, pp. 100–111, 2009
318. R. S. Williams, G. E. Dodson, O. Limbo, et al., "Nbs1 is an extended flexible arm binding to Ctp1 and Mre11-Rad50 to coordinate dsDNA break processing," Cell, vol. 139, pp. 87–99, 2009.
319. T. T. Paull and M. Gellert, "A mechanistic basis for Mre11-directed DNA joining at microhomologies,"Proceedings of the National Academy of Sciences of the United States of America, vol. 97, no. 12, pp. 6409–6414, 2000
320. L. Chen, K. Trujillo, W. Ramos, P. Sung, and A. E. Tomkinson, "Promotion of Dnl4-Catalyzed DNA end-joining by the Rad50/Mre11/Xrs2 and Hdf1/Hdf2 complexes," Molecular Cell, vol. 8, no. 5, pp. 1105–1115, 2001
321. L. Aravind, V. M. Dixit, and E. V. Koonin, "Apoptotic molecular machinery: vastly increased complexity in vertebrates revealed by genome comparisons," Science, vol. 291, no. 5507, pp. 1279–1284, 2001

322. M.-P. Doutriaux, F. Couteau, C. Bergounioux, and C. White, "Isolation and characterisation of the RAD51 and DMC1 homologs from Arabidopsis thaliana," Molecular and General Genetics, vol. 257, no. 3, pp. 283–291, 1998
323. A. C. Vergunst and P. J. J. Hooykaas, "Recombination in the plant genome and its application in biotechnology," Critical Reviews in Plant Sciences, vol. 18, no. 1, pp. 1–31, 1999
324. M. Majchrzak, R. P. Bowater, P. Staczek, and P. Parniewski, "SOS repair and supercoiling influence the genetic stability of DNA triplet repeats in Escherichia coli," Journal of Molecular Biology, vol. 364, no. 4, pp. 612–624, 2006
325. J. Courcelle, A. Khodursky, B. Peter, P. O. Brown, and P. C. Hanawalt, "Comparative gene expression profiles following UV exposure in wild-type and SOS-deficient Escherichia coli," Genetics, vol. 58, pp. 41–64, 2001.
326. B. Michel, "After 30 years of study, the bacterial SOS response still surprises us," PLoS Biology, vol. 3, no. 7, pp. 1174–1176, 2005
327. M. Martins-Pinheiro, R. C. P. Marques, and C. F. M. Menck, "Genome analysis of DNA repair genes in the alpha proteobacterium Caulobacter crescentus," BMC Microbiology, vol. 7, article no. 17, 2007
328. B. Van Houten, J. A. Eisen, and P. C. Hanawalt, "A cut above: discovery of an alternative excision repair pathway in bacteria," Proceedings of the National Academy of Sciences of the United States of America, vol. 99, no. 5, pp. 2581–2583, 2002
329. R. P. Fuchs, S. Fujii, and J. Wagner, "Properties and functions of Escherichia coli Pol IV and Pol V,"Advances in Protein Chemistry, vol. 69, pp. 229–264, 2004
330. M. F. Goodman, "Coping with replication ‹train wrecks› in Escherichia coli using Pol V, Pol II and RecA proteins," Trends in Biochemical Sciences, vol. 25, no. 4, pp. 189–195, 2000
331. B. S. Strauss, R. Roberts, L. Francis, and P. Pouryazdanparast, "Role of the dinB gene product in spontaneous mutation in Escherichia coli with an impaired replicative polymerase," Journal of Bacteriology, vol. 182, no. 23, pp. 6742–6750, 2000

332. R. S. Galhardo, R. P. Rocha, M. V. Marques, and C. F. M. Menck, "An SOS-regulated operon involved in damage-inducible mutagenesis in Caulobacter crescentus," Nucleic Acids Research, vol. 33, no. 8, pp. 2603–2614, 2005
333. L. H. Hartwell and T. A. Weinert, "Checkpoints: controls that ensure the order of cell cycle events,"Science, vol. 246, no. 4930, pp. 629–634, 1989.
334. H.-L. Lo, S. Nakajima, L. Ma et al., "Differential biologic effects of CPD and 6–4PP UV-induced DNA damage on the induction of apoptosis and cell-cycle arrest," BMC Cancer, vol. 5, article no. 135, 2005
335. J. Mercer, M. Mahmoudi, and M. Bennett, "DNA damage, p^{53}, apoptosis and vascular disease,"Mutation Research, vol. 621, no. 1-2, pp. 75–86, 2007
336. J. Lukas, C. Lukas, and J. Bartek, "Mammalian cell cycle checkpoints: signalling pathways and their organization in space and time," DNA Repair, vol. 3, no. 8-9, pp. 997–1007, 2004
337. Y. Shiloh, "ATM and related protein kinases: safeguarding genome integrity," Nature Reviews Cancer, vol. 3, no. 3, pp. 155–168, 2003
338. P. A. Jeggo and M. Löbrich, "Contribution of DNA repair and cell cycle checkpoint arrest to the maintenance of genomic stability," DNA Repair, vol. 5, no. 9-10, pp. 1192–1198, 2006
339. S. S. Zinkel, K. E. Hurov, C. Ong, F. M. Abtahi, A. Gross, and S. J. Korsmeyer, "A role for proapoptotic BID in the DNA-damage response," Cell, vol. 122, no. 4, pp. 579–591, 2005
340. I. Kamer, R. Sarig, Y. Zaltsman et al., "Proapoptotic BID is an ATM effector in the DNA-damage response," Cell, vol. 122, no. 4, pp. 593–603, 2005
341. J. Jin, T. Shirogane, L. Xu et al., "SCFβ-TRCP links Chk1 signaling to degradation of the Cdc25A protein phosphatase," Genes and Development, vol. 17, no. 24, pp. 3062–3074, 2003
342. Y. Kanemori, K. Uto, and N. Sagata, "β-TrCP recognizes a previously undescribed nonphosphorylated destruction motif in Cdc25A and Cdc25B phosphatases," Proceedings of the National Academy of Sciences of the United States of America, vol. 102, no. 18, pp. 6279–6284, 2005

343. N. Mailand, J. Falck, C. Lukas et al., "Rapid destruction of human Cdc25A in response to DNA damage,"Science, vol. 288, no. 5470, pp. 1425–1429, 2000.
344. N. Mailand, A. V. Podtelejnikov, A. Groth, M. Mann, J. Bartek, and J. Lukas, "Regulation of G_2/M events by Cdc25A through phosphorylation-dependent modulation of its stability," EMBO Journal, vol. 21, no. 21, pp. 5911–5920, 2002.
345. J. Bartek and J. Lukas, "DNA damage checkpoints: from initiation to recovery or adaptation," Current Opinion in Cell Biology, vol. 19, no. 2, pp. 238–245, 2007
346. N. Watanabe, H. Arai, Y. Nishihara et al., "M-phase kinases induce phospho-dependent ubiquitination of somatic Wee1 by SCFβ-TrCP," Proceedings of the National Academy of Sciences of the United States of America, vol. 101, no. 13, pp. 4419–4424, 2004
347. N. Mailand, S. Bekker-Jensen, J. Bartek, and J. Lukas, "Destruction of claspin by SCFβTrCP restrains Chk1 activation and facilitates recovery from genotoxic stress," Molecular Cell, vol. 23, no. 3, pp. 307–318, 2006
348. A. Peschiaroli, N. V. Dorrello, D. Guardavaccaro et al., "SCFβTrCP-mediated degradation of claspin regulates recovery from the DNA replication checkpoint response," Molecular Cell, vol. 23, no. 3, pp. 319–329, 2006
349. I. Mamely, M. A. van Vugt, V. A. Smits et al., "Polo-like kinase-1 controls proteasome-dependent degradation of Claspin during checkpoint recovery," Current Biology, vol. 16, no. 19, pp. 1950–1955, 2006
350. L. Busino, M. Donzelli, M. Chiesa et al., "Degradation of Cdc25A by β-TrCP during S phase and in response to DNA damage," Nature, vol. 426, no. 6962, pp. 87–91, 2003
351. M. B. Kastan and J. Bartek, "Cell-cycle checkpoints and cancer," Nature, vol. 432, no. 7015, pp. 316–323, 2004
352. C. Masutani, R. Kusumoto, S. Iwai, and F. Hanaoka, "Mechanisms of accurate translesion synthesis by human DNA polymerase η," EMBO Journal, vol. 19, no. 12, pp. 3100–3109, 2000.
353. R. E. Johnson, S. Prakash, and L. Prakash, "Efficient bypass of a thymine—thymine dimer by yeast DNA polymerase, Polη," Science, vol. 283, no. 5404, pp. 1001–1004, 1999

354. R. P. Rastogi, Richa, and R. P. Sinha, "Apoptosis: molecular mechanisms and pathogenicity," EXCLI Journal, vol. 8, pp. 155–181, 2009.

Chapter 6

Rates of Molecular Evolution and Diversification in Plants: chloroplast Substitution Rates Correlate with Species-Richness in the Proteaceae

David Duchene and Lindell Bromham

Centre for Macroevolution and Macroecology, Evolution, Ecology & Genetics, Research School of Biology, Australian National University, Canberra, ACT, 0200, Australia

ABSTRACT

Background

Many factors have been identified as correlates of the rate of molecular evolution, such as body size and generation length. Analysis of many molecular phylogenies has also revealed correlations between substitution rates and clade size, suggesting a link between rates of molecular evolution and the process of diversification. However, it is not known whether this relationship applies to all lineages and all sequences. Here, in order to investigate how widespread this phenomenon is, we investigate patterns of substitution in chloroplast genomes of the diverse angiosperm family Proteaceae. We used DNA sequences from six chloroplast genes (6278bp alignment with 62 taxa) to test for a correlation between diversification and the rate of substitutions.

Results

Using phylogenetically-independent sister pairs, we show that species-rich lineages of Proteaceae tend to have significantly higher chloroplast substitution rates, for both synonymous and non-synonymous substitutions.

Conclusions

We show that the rate of molecular evolution in chloroplast genomes is correlated with net diversification rates in this large plant family. We discuss the possible causes of this relationship, including molecular evolution driving diversification, speciation increasing the rate of substitutions, or a third factor causing an indirect link between molecular and diversification rates. The link between the synonymous substitution rate and clade size is consistent with a role for the mutation rate of chloroplasts driving the speed of reproductive isolation. We find no significant differences in the ratio of non-synonymous to synonymous substitutions between lineages differing in net diversification rate, therefore we detect no signal of population size changes or alteration in selection pressures that might be causing this relationship.

BACKGROUND

Present biodiversity has come about through processes of diversification and extinction of species, and the mechanisms that drive these processes are a central focus in evolutionary biology (e.g.[1-4]). One intriguing relationship that has been revealed through studies of branch lengths on molecular phylogenies is a link between the rate of molecular evolution and the net diversification rate. A correlation between evolutionary rates and species diversity has been found in several groups including flowering plants [5,6], reptiles [7], birds [7,8], and other metazoan phyla, orders, and classes [9].

Table 1: Genera of sister clades, and their corresponding species richness and branch length estimates

Pair	Sister clades	Taxa	Species richness	Genes available	Total branch lengths	dN branch lengths	dS branch lengths	dN/dS
1	Persoonia	*Persoonia spp.*	100	2	0.00303	0.00331	0.00299	1.10704
	Garnieria + Acidonia	*Garnieria spathulaefolia*	2		0.00097	0.00105	0.00170	0.61562
2	Symphionema	*Symphionema montanum*	2	4	0.01329	0.00990	0.01713	0.57818
	Agastachys	*Agastachys odorata*	1		0.00828	0.00403	0.02123	0.18971
3	Cenarrhenes + Dilobeia	*Cenarrhenes nitida*	3	1	0.01703	0.00379	0.02849	0.13315
	Beaupreopsis	*Beaupreopsis paniculata*	1		0.00936	0.00315	0.01317	0.23892
4	Conospermum + Synaphea	*Conospermum spp.*	104	4	0.03480	0.01312	0.06744	0.19454
	Stirlingia	*Stirlingia latifolia*	7		0.02865	0.01164	0.05841	0.19922
5	Protea	*Protea cynaroides*	112	5	0.00872	0.00588	0.01364	0.43081
	Faurea	*Faurea spp.*	15		0.00660	0.00589	0.01052	0.55959
6	Petrophile	*Petrophile spp.*	53	3	0.01424	0.00723	0.02834	0.25512
	Aulax	*Aulax spp.*	3		0.01537	0.01279	0.02951	0.43337
7	Paranomus	*Paranomus spp.*	19	2	0.00130	0.00187	1.00E-09	1.87E + 06
	Vexatorella	*Vexatorella alpina*	4		0.00130	0.00047	0.00445	0.10504
8	Leucospermum	*Leucospermum spp.*	48	2	0.00129	0.00095	0.00292	0.32577
	Orothamnus + Diastella + Mimetes	*Mimetes spp.*	21		0.00183	0.00187	0.00296	0.63319
9	Alloxylon + Oreocallis	*Alloxylon spp.*	6	4	0.00641	0.00418	0.00931	0.44871
	Embothrium	*Embothrium coccineum*	1		0.00577	0.00323	0.00625	0.51730

10	Stenocarpus + Strangea	*Stenocarpus salignus*	26	6	0.01094	0.00602	0.01962	0.30693
	Lomatia	*Lomatia spp.*	12		0.00625	0.00395	0.01304	0.30245
11	Grevillea + Finschia + Hakea	*Grevillea spp.*	515	6	0.01847	0.01027	0.03031	0.33881
	Buckinghamia	*Buckinghamia spp.*	2		0.00306	0.00118	0.00916	0.12925
12	Virotia	*Virotia leptophylla*	6	1	0.00091	0.00097	1.00E-09	9.71E + 05
	Athertonia	*Athertonia diversifolia*	1		0.00754	0.00535	0.01243	0.43028
13	Panopsis + Brabejum	*Panopsis spp.*	27	4	0.00966	0.00577	0.01995	0.28920
	Macadamia	*Macadamia spp.*	9		0.00752	0.00623	0.01371	0.45473
14	Hicksbeachia	*Hicksbeachia pinnatifolia*	2	1	0.00229	0.00261	0.00095	2.75046
	Gevuina	*Gevuina avellana*	1		0.00220	0.00098	0.00309	0.31894
15	Euplassa	*Euplassas occidentalis*	20	3	0.00318	0.00281	0.00448	0.62802
	Sleumerodendron + Kermadecia + Turrillia	*Sleumerodendron austrocaledonicum + Kermadecia pronyensis*	8		0.00262	0.00199	0.00619	0.32064
16	Banksia + Dryandra	*Banksia spp.*	169	6	0.01403	0.00893	0.02906	0.30745
	Austromuellera + Musgravea	*Austromuellera trinervia*	4		0.01107	0.00525	0.02498	0.21017
17	Roupala + Neorites	*Roupala montana + Roupala monosperma + Neorites kevediana*	34	6	0.00620	0.00286	0.01367	0.20891
	Orites	*Orites spp.*	8		0.00420	0.00229	0.00483	0.47439
18	Darlingia	*Darlingia darlingiana*	2	2	0.00219	0.00236	0.00295	0.79752
	Floydia	*Floydia praealta*	1		0.00297	0.00132	0.00571	0.23042
19	Lambertia	*Lambertia spp.*	10	4	0.02536	0.01116	0.05726	0.19491
	Xylomelum	*Xylomelum spp.*	6		0.02419	0.01036	0.04448	0.23296
20	Helicia	*Helicia spp.*	97	3	0.00149	0.00094	0.00304	3.10E-01
	Hollandaea	*Hollandaea riparia*	4		0.00039	0.00000	0	0

Detailed legend: Sister comparisons included in this analysis. For each pair, we list the two clades compared (a "+" indicates where more than one genus were combined as one sister lineage of a comparison). The taxon from which the sequence was taken is listed, with "spp" indicating that several congeneric sequences were combined (see Table S1 for details). The species richness for each sister lineage is taken from Sauquet et al. (2009), and the number of available gene sequences for each comparison, out of the six chloroplast genes analysed in this study The estimated substitution rates for each sister lineage are given for Total branch lengths (all substitutions) as well as synonymous (dS) and non-synonymous (dN) substitutions, and estimates of **ω** (dN/dS).

Duchene and Bromham

Duchene and Bromham *BMC Evolutionary Biology* 2013 13:65, doi: 10.1186/1471-2148-13-65

However, not all the datasets analysed have provided evidence for a link between diversification rates and rates of molecular evolution. One study of a large number of phylogenies found a relationship between root-to-tip branch lengths and net diversification in around half of the phylogenies tested, but it is not clear whether this was due to low power or lack of a common trend [10]. Another study examined genetic data within the mammals and found no evidence of an association between molecular rates and net diversification [11]. Accordingly, the universality and causes of the link remain uncertain.

There are three possible causes for the association between net diversification and the rate of molecular evolution [5,6]. One is that the process of diversification drives changes in the rate of molecular evolution. Speciation might influence the rate of molecular evolution through positive selection on particular genes associated with adaptation to novel niches [12]. Speciation could also cause genome-wide increases in substitution rate if speciation is typically associated with population subdivision [13,14]. This is because a reduction in effective population size (*Ne*) can cause a higher rate of fixation of nearly neutral mutations (e.g. [15]), leading to a faster substitution rate [14].

Conversely, a higher rate of molecular evolution may increase the diversification rate. A faster mutation rate may hasten differentiation between populations and promote reproductive incompatibility [16]. For example, it has been suggested that higher standing genetic diversity in populations at low latitude may contribute to faster diversification in the tropics [17]. Increased standing genetic variation may produce more raw material for adaptation [18] or reduce the likelihood of extinction [19]. However, a recent study of orchids found no evidence for a link between population genetic variability and net diversification rate [20]. A higher rate of molecular evolution may increase the rate of diversification by accelerating the formation of hybrid incompatibility, occurring through the accumulation of genetic incompatibilities between the genomes of the diverging populations [16].

Alternatively, there might be a third factor that influences both the rate of molecular evolution and diversification rate, creating an indirect link between diversification and molecular evolution. For example, environmental energy (temperature and UV light) has been associated with both the diversification rate and the rate of molecular evolution in angiosperms [21]. Other potential third factors are life history features, such as size or generation length, which are linked with the rate of molecular evolution and diversification rates of angiosperms and several metazoan taxa [9,22-24]. It has also been suggested that both morphological and molecular rates of change may be connected to diversification rate [24]. Whether the correlation between rate of molecular evolution and net diversification has a causal or indirect effect needs more investigation.

One way of disentangling the potential causes of the observed relationship between diversification rate and rate of molecular evolution is to partition substitutions in protein-coding genes into synonymous and non-synonymous substitutions. Synonymous mutations do not change the amino acid sequence of a protein and hence are expected to behave as neutral. If so, then the synonymous substitution rate (dS) should reflect only the mutation rate [25]. Nonsynonymous mutations are expected to have a range of fitness effects, including neutral, positive and negative, so may be subject to both drift and selection. An increase in the nonsynonymous substitution rate (dN) relative to the synonymous rate (dS) can occur through positive selection promoting the fixation of nonsynonymous mutations, or through a reduction in population size increasing the rate of fixation of nearly neutral mutations by drift.

The link between rate of molecular evolution and diversification rate has been attributed to the action of selection during speciation, or to a reduction in average population size in lineages undergoing diversification [14], both of which would be expected to increase the relative rate of nonsynonymous substitutions. However, studies in angiosperms [6], reptiles [7], and birds [7,8] have found a correlation between synonymous substitutions and net diversification, leading to the suggestion that the link between molecular rates and net diversification may be driven by the mutation rate.

Here, we focus on the rate of molecular evolution in chloroplast genes. Genetic changes in chloroplast genomes have been implicated in the process of speciation in plants. Coevolution between organelle and nuclear genomes has been recognized as an important factor in plant diversification [26]. Plastome-genome incompatibility can cause hybrid sterility or inviability, by disrupting sexual reproduction, leaf morphologies, and machineries for photosynthesis or respiration [27-29]. Some of the genetic events in chloroplasts that produce these aberrations are gene duplications, loss of gene complexes and genome rearrangements [26, 30, 31]. The resulting incompatibilities are probably generalized phenomena in plants, and the evolutionary consequence is that they can enhance post-zygotic barriers during speciation [26, 29, 32, 33]. It seems possible, then, that variation in rates of molecular evolution of chloroplasts could also influence the speed of genetic isolation, and hence the diversification rate of plant lineages.

Using a phylogenetic comparative analysis of sister pairs [34], we investigated the relationship between rates of molecular evolution and net diversification in chloroplast genes of the plant family Proteaceae. This highly diverse family is mostly restricted to the Southern Hemisphere. It contains 79 recognized genera and around 1600 species, and some of its most diverse groups are the Australian genus *Banksia* and the African genus *Protea*. The high diversity of Proteaceae makes it a particularly attractive case study for diversification (e.g. [35-37]). In addition, the family has stark contrasts in species-richness between genera even within its biodiversity hotspots [38]. Of particular interest to this study are the numerous cases of monophyletic sister clades with remarkable differences in number of species. For example, the genus *Protea* has 112 species, while its sister genus *Faurea* has 15, and the *Banksia* lineage (including the dryandras) has 169 species while its sister lineage of the genera *Austromuellera* and *Musgravea* contains only 4.

We focus on the rates of evolution of six chloroplast genes available for a genus level phylogeny of the family Proteaceae [38]. We use three protein-coding genes to estimate and contrast rates of synonymous (dS) and non-synonymous (dN) substitutions. Comparing dN, dS, and ω (dN/dS) to species-richness of clades allows us to separate the effect of mutation rate on net diversification from the effect of selection and effective population size. In this way, we aim to provide insight into the factors underlying the correlation between rates of molecular evolution and net diversification.

RESULTS

A model where every branch in the phylogeny had an independent rate of substitutions had a significantly higher likelihood than the constant rates model in all the rates estimations (all substitutions, dN, dS), and ω (P value < 0.01 for all tests; see Methods section), indicating that the rate of molecular evolution of the chloroplast genes analysed varies significantly between lineages of the family Proteaceae.

Species-rich lineages had significantly longer branch lengths in the phylogeny estimated from the full 6-gene dataset (one-tailed Wilcoxon Signed-Rank test, $W = 175$, $P = 0.0036$). This is evidence of a positive association between net diversification and the rate of molecular evolution of chloroplasts in the family Proteaceae. We also found significant differences in estimates of synonymous (dS: $W = 152$, $P = 0.041$), and non-synonymous rates (dN: $W = 165$, $P = 0.012$). However, we did not find a significant differences in estimates of ω between species-rich and species-poor sister lineages ($W = 100$, $P = 0.14$).

DISCUSSION

We found a significant positive association between the rate of molecular evolution in chloroplast genes and species-richness in the plant family Proteaceae. There were significant associations between both synonymous and non-synonymous rates of substitutions and net diversification, but not between ω (dN/dS) and diversification. The pattern of correlations in this study are consistent with other studies of angiosperms [6, 39], reptiles, and birds [7, 8, 10]. Importantly, our results give some insight into the cause of this relationship. The variation in both synonymous and non-synonymous substitution rates between lineages may reflect a role for the rate of production of mutations in the chloroplast genome in the process of diversification in the Proteaceae. Because we fail to detect an increase in ω in species-rich clades, our analysis provides no tangible evidence for a role of selection or population size change in driving the relationship between substitution rates and diversification rates in this group.

Synonymous Substitutions and Net Diversification

Synonymous substitution rates are typically interpreted to reflect the rate of production of mutations. Mutation rates are known to vary between lineages for a range of reasons. For example, species with shorter generation times tend to have faster mutation rates [23], presumably due to the accumulation of DNA replication errors [40]. Mutation rates can also vary across the genome, which may be at least in part due to differences in base composition or gene length [24, 41].

Since synonymous substitutions are commonly assumed to be functionally neutral, they are often used to provide a window into variation of mutation rates. However, bias in codon use can influence the synonymous rate if, for example, there is selection for efficiency in the process of translation [42]. This type of bias has been found in angiosperm mitochondrial genes although with selection that is so weak that it is considered not to affect estimations of mutation rates [43]. The chloroplast genome of angiosperms also has minimal codon bias and weak selection for translation efficiency [44]. Therefore, in this study, we consider that the relationship between synonymous substitution rate and net diversification is telling us something about the link between mutation and diversification, whether it reflects differences in the absolute mutation rate per unit time or in the differences in the distribution of fitness effects of synonymous mutations between lineages.

One explanation for the link between synonymous substitutions and net diversification is that higher mutation rates could cause faster genetic divergence between lineages. In this case, genes of chloroplast origin may be important because they can drive reproductive isolation in plants by interacting with nuclear alleles [25]. Reproductive barriers can occur due to the failure of interactions between nuclear and cytoplasmic gene complexes, for example cytoplasmic male sterility [45]. An increased mutation rate may generate more molecular changes that cause these phenomena, known as Bateson-Dobzhansky-Muller (BDM) incompatibilities, and so might accelerate post-zygotic isolation [16,46,47].

Some studies have found that lineage-specific variation in rates of molecular evolution are consistent across the nuclear, mitochondrial, and chloroplast genomes [48], so it may be that the increase in substitution rates that we detected also apply to the nuclear genomes of species-rich lineages in the Proteaceae. In this case, higher rates of substitution in the nuclear genome may be contributing to the formation of incompatibilities between diverging populations, either by generating BDM incompatibilities between the nuclear genomes or through interactions between the nuclear and organelle genomes.

Therefore, the association between the synonymous rate of chloroplast genes and diversification rate reported here may reflect the acceleration in the formation of post-zygotic reproductive isolation. This is also consistent with our finding of an association between non-synonymous rates and net diversification because an increase in the mutation rate should also result in more effectively neutral non-synonymous substitutions going to fixation.

Indirect Links between Diversification and the Rate of Molecular Evolution

An indirect relationship between the rate of molecular evolution and diversification could arise if some factor influenced both. For example, it has been suggested that tropical lineages have a higher rate of molecular evolution than their temperate counterparts [49]. This correlation might reflect a direct effect of temperature or UV light on mutagenesis [50], or an indirect effect if higher environmental energy leads to further growth rates and more rapid generation turnover, which could influence the mutation rate through accumulation of replication errors [23]. If higher growth rates also lead to faster diversification [51], then this could create an indirect link between the mutation rate and diversification. This may also explain the patterns in a study on angiosperms that investigated the correlations between species-richness, the rate of molecular evolution, and three energy variables (temperature, UV light, and evapotransportation), but which found no support for the mutation rate as the direct mediator of species-richness [21]. However, it is interesting to note that the Proteaceae do not appear to have higher rates of diversification in the tropics. Instead, much of their radiation has occurred in Mediterranean climate hotspots [38].

Life history variation provides another possible indirect link between rates of molecular evolution and diversification. Several studies have suggested that annual plants have a faster rate of molecular evolution than perennials, a pattern generally attributed to the generation time effect (see [24,40]). The potential for interactions between mechanisms that influence species-richness and the rate of molecular evolution has a broad scope and remains to be studied in detail.

Net Diversification and

It has been suggested that processes associated with speciation drive the link between rates of substitution and net diversification [10, 13, 14], including diversifying selection and changes in effective population size. A reduction in effective population size (*Ne*) may be caused by a speciation event that changes the population structure, such as vicariant or peripatric speciation [52]. This could lead to new adaptive pressures [53], or high levels of genetic drift in population bottlenecks [54]. These processes could increase the rate of fixation on non-synonymous substitutions, which may be reflected in an increase in ω (dN/dS) [52, 55, 56].

Two studies on large numbers of phylogenies found a recurrent correlation between root-to-tip distances and the number of speciation events [10, 13]. This result was interpreted as evidence that clades with more speciation events have a faster rate of molecular evolution, which they attributed to punctuational change associated with the founder-effect model of speciation. However, while these phylogenetic tests reveal an association between rates of evolution and number of phylogenetic nodes, they are not able to localise those changes to the nodes rather than the edges of the phylogeny, so cannot distinguish between two alternative explanations, that speciation events increase the substitution rate or that higher substitution rate increases diversification. One possible way to separate these models is in their predicted effects on the patterns of substitutions. If population divisions associated with speciation events have significant effects on rates of substitution, either through change in selection or reduction in effective population size, it should result in a relative increase in the nonsynonymous rate, reflected in an increase in dN/dS (ω).

We did not detect any association between ω and net diversification (see also [8]). This may be because net diversification is not associated with consistent effects on population size [46], or diversification does affect effective population size, but the effect on ω is overwritten by other population fluctuations [46]. Alternatively, the effect on reduction in effective population size may be too small to be detected or may be affected by the method of estimation of ω [57]. In theory, *Ne* is an adequate representation of genetic drift in large populations and when the population size has been consistent for a long enough time [52]. It has even been shown that following transient increases in *Ne* there can be an increase in the rate of substitutions due to slightly advantageous mutations, which is the opposite of the predicted effect [58]. Therefore, although *Ne* is likely to have a significant effect on the rate of substitutions, predicting the form of the effect is far from a simple task [56]. Therefore, failure to detect an effect of population size changes of ω in this study does not imply that *Ne* is unaffected by diversification; however, it does suggest that changes in *Ne*during diversification are unlikely to explain the differences in substitution rates that we observe in these data.

Molecular Evolution and Diversification in Plants

Many studies have focussed on identifying the genetic loci underlying speciation. These can be genes that contribute to the genetic isolation of populations, genes that drive differential ecological adaptation, and "magic traits" that do both (e.g. [59, 60]). Genome-wide scans are increasingly being used to identify outlier loci that show signatures of selection, including loci that differ between pairs that are associated with floral traits, climatic factors, and sterility [61]. This study takes a different, and complementary, approach to analysing the role that the genome-wide generation of genetic change plays in diversification.

It is possible that higher mutation rates may create a greater pool of standing variation from which adaptive substitutions can be derived. The assumption that chloroplast genes do not play a direct role in ecological adaptation has now been challenged: for example, values of ω above 1 have been estimated in rbcL [62] and MatK sequences [63] for some linages, which was interpreted as a signal of positive selection. However, we did not find evidence of higher ω in more diverse clades for the loci analysed in this study, and the relationship between the amount of standing variation and diversification in plants is not clear. For example, studies have found that diversification in orchids is not associated with greater genetic diversity at the population level [20, 64].

Another scenario is that higher mutation rates contribute to the rate at which the genomes of different populations diverge and become gradually incompatible, making hybrids between the populations less fit. Bateson-Dozhansky-Muller (BDM) incompatibilities may arise from selection in different populations, but they might also be unconnected to ecological or behavioural divergence, in other words they may be "incidental on other acquired differences" [65]. For a mutation to go to substitution in one population, it must be broadly compatible with other common alleles in that population. But it will not have been "tested by natural selection" against alleles in isolated populations, and bringing those unharmonised alleles together may result in a maladapted individual [66]. The more unique substitutions each population has acquired, the greater the chance that a hybrid zygote will contain at least one pair of incompatible alleles. The steady increase in hybrid incompatibility with time in many species has been taken as evidence that many loci may contribute to BDM incompatibilities [47]. Under the BDM model, the rate of speciation may increase as the mutation rate increases [67]. Since the substitutions underlying BDM incompatibilities do not have to occur evenly in both lineages, a higher mutation rate in one lineage should drive divergence between them [47]. Debate continues over the rate at which hybrid incompatibility accumulates, particularly concerning the prediction that BDM incompatibilities should "snowball", accelerating relative to the substitution rate [68,69].

Importantly, incompatibilities between populations can involve both organelle and nuclear genomes. Just as alleles within the nuclear genome must be able to work together to produce viable offspring, genomes of chloroplasts must be co-adapted to nuclear genome to allow normal development [70]. For example, alleles that cause cytoplasmic male sterility may be countered by suppression genes in the nuclear genome that restore male function, so a hybrid that inherited the organelle genome without the corresponding nuclear allele would be male sterile [33]. While cytonuclear conflict has been more frequently studied between mitochondrial and nuclear genomes, there is increasing evidence that incompatibilities between chloroplast and nuclear genomes contribute to hybrid incompatibility in many plant species [26].

Polyploidy is another important factor in the diversification of many plant lineages [71, 72], but by focussing only on the chloroplast we minimized the impact of genome duplication on our analyses. Chloroplasts typically have uniparental inheritance, which simplifies the interpretation of the effects of genetic changes on divergence. However, the mode of inheritance of chloroplasts, whether inherited paternally or maternally, can vary between taxa, which can influence their levels of genetic diversity [73]. Chloroplast sequences should also limit the impact of "divergence hitchhiking", where linked neutral loci go to fixation through being linked to a locus under selection [61]. Lastly, while chloroplasts use recombination for genome repair [74], hybridization of chloroplasts from different lineages does not appear to be common [26, 75].

CONCLUSIONS

We show a significantly faster rate of molecular evolution in chloroplast genes of species-rich lineages of the family Proteaceae. These results offer evidence for the influence of the rate of molecular evolution on diversification. This does not imply that the rate of molecular evolution explains the process of diversification, because this complex and heterogeneous process can be influenced by many mechanisms such as hybridization [76], polyploidy [77], allopatric events [14], and the duplication of genes [30]. However, the results do suggest that the substitution rate in chloroplasts may be one of these influences on the speed at which populations diverge, thus influencing the probability of populations becoming separate species [16].

METHODS

Sister Pairs

The phylogenetic analysis of the family Proteaceae by Sauquet et al. (2009) includes the 79 recognized genera and the species-richness for each genus compiled from the literature (references are also available in [78]). For the present study, monophyletic pairs of sister lineages that display differences in current species-richness were chosen from Sauquet et al's phylogeny. The main criterion to select pairs was that the pair was monophyletic, so the two sister lineages had the same amount of time to accumulate species diversity and substitutions, and each sister pair was phylogenetically independent from all other such pairs.

We chose one genus to represent each sister lineage in order to avoid bias in branch length estimation due to the node density effect [79]. The chosen genus was the one with the greatest gene coverage. If the genera of a sister lineage had equal genetic coverage the genus was chosen at random. Using only one sequence per genus may reduce the power of the test, which may obscure a weak pattern; but using only one randomly selected species per sister lineage is unlikely to generate any systematic biases in rates, making this approach conservative for testing an association between rates of molecular evolution and net diversification (see [8,11]). In some cases a sister lineage is represented by combining sequences from several closely related genera. This practice increases our power to resolve the shared history of that lineage since its divergence from the common ancestor of the sister pair, and this is unlikely to create any systematic biases in rate estimations.

Molecular Dataset

Branch length estimation was critical for comparative analyses in this study, so the genetic dataset required unambiguous genetic alignments and the maximum gene coverage of the species analysed. With these criteria we included six genes of chloroplast origin (*atpB*, *atpB-rbcL*, *matK*, *rbcL*, *trnL* intron, and *trnL-trnF*) from the data by Sauquet et al. 2009 and available in the GenBank repository. These were then aligned using the MUSCLE algorithm, checked by eye, and manually corrected using the program SeaView v4 [80]. This resulted in a 6278bp alignment with 62 taxa, 4457bp of exons, and 1821bp of introns.

Phylogenetic Estimation

Each gene alignment was tested for the most appropriate model of substitutions using likelihood estimation and comparison with the Bayesian Information Criterion as implemented in the package "ape" [81] in R (The R Project - http://www.r-project.org/ *webcite*). Applying a partition by genes with the models selected a maximum likelihood analysis with 1000 replicates was run using Garli v2.0 [82]. The resulting tree was then used to extract the branch length values of the sister pairs (Figure 1). If any of our chosen sister pairs were not monophyletic in our phylogeny they were excluded from the analysis. Twenty sister pairs of the initial twenty-two chosen from Sauquet at al's phylogeny were monophyletic in our estimates (Figure 1).

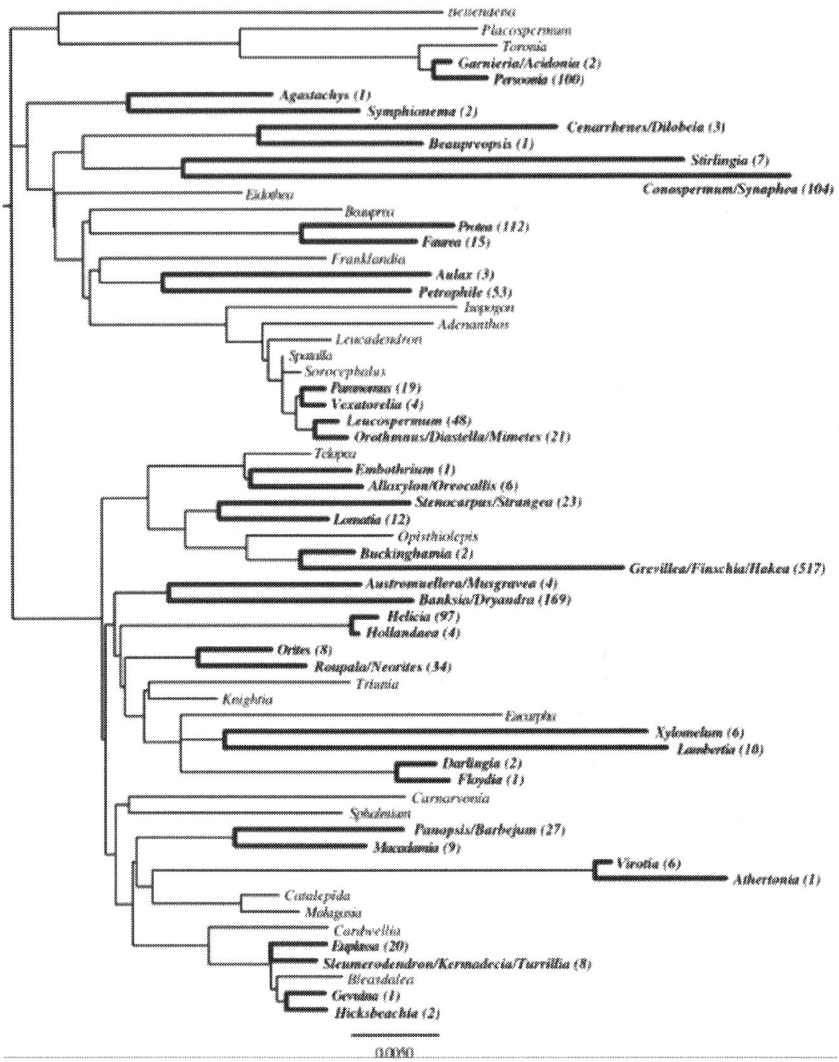

Figure 1: Phylogeny of Proteaceae with the sister pairs analysed in bold. Detailed legend: Molecular phylogeny of the family Proteaceae including the 20 sister pair groups used for the present analyses in bold. The branch lengths shown are proportional to the estimated number of substitutions as indicated by the whole dataset of six chloroplast genes. The species richness of the clades compared is shown in parentheses, with species numbers taken from Sauquet et al. (2009; references are also available in [78]). The scale bar indicates number of mutations.

The inferred phylogenies (including those inferred for synonymous and non-synonymous substitutions; see next section) were examined for significant overall variation in branch lengths. To do this, we estimated the likelihoods of both a constant rates model and a free rates model (where there is one rate per branch) in the program HyPhy v2.1 [83], and compared them using a likelihood ratio test. If the free rates model provides a better fit for the data, this suggests significant variation in rates of molecular evolution across the phylogeny.

dN and dS Trees

To examine the potential link between synonymous substitutions (dS), non-synonymous substitutions (dN), and ω (dN/dS) and species-richness, we estimated dN and dS rates using an alignment of the coding genes (*atpB*, *matK*, and *rbcL*) in the program HyPhy v2.1 [83] with the MG94 model of codon evolution [84]. For the estimation of dN and dS trees, the MG94 model can be combined with any of the nucleotide substitution models nested in GTR + G + I. To choose the best combination we first gave HyPhy v2.0 a notation to estimate the codon frequencies (the frequency of each of the four bases in each of the three codon positions), which was a 3x4 matrix. Then the likelihood of each of the 203 possibilities was estimated and one was chosen according to the Akaike Information Criterion. The model chosen had four parameters, where $\theta_{AG} = \theta_{CT}$ and $\theta_{CG} = \theta_{GT}$, and its label in HyPhy v2.0 is MG94_3x4_012313. Finally, these parameters were optimized with maximum likelihood, constraining the topology to that estimated from the full six-gene dataset (Figure 1). The output included the dN and dS branch lengths, which were used to extract the branch lengths of sister pairs. Given that HyPhy v2.0 estimates the values of dN and dS as the expected number of substitutions per nucleotide per site, the values for ω were calculated as the ratio between the two estimates

Statistics

The total species-richness and the estimates of branch lengths (for all substitutions, dN and dS) and ω were collected for each of the two lineages in the sister pairs (Table 1). As the sister lineages had the same amount of time to accumulate species and substitutions, we assumed that the branch length is proportional to the rate of molecular evolution of chloroplasts (reviewed in [34]). Similarly, we assumed that species-richness of each sister clade reflects the net diversification (speciation minus extinction) of that lineage since the last common ancestor of the sister pair.

We performed a one-tailed Wilcoxon Signed-Ranks test in R, which resembles the standard sign test but accounts for the magnitudes of the differences between matched lineages [85]. This test sets a sign to each pair by subtracting branch lengths in the direction from species-rich to species poor; we did not include the sister pairs with equal species-richness as these cannot be accommodated in the Wilcoxon Signed-Ranks test. Then, the absolute difference between the two values was used to rank the pairs (lowest difference has rank 1 and the highest rank is the number of pairs). Tied values receive as a rank the mean of the ranks they span. The ranks are then given the sign of the pair and then added to produce a W statistic [85].

AUTHORS' CONTRIBUTIONS

David Duchene: ES & FG. Lindell Bromham: ES & FG. ES: Participated in the design of the study and performed statistical analyses. FG: conceived of the study, and participated in its design and coordination and helped to draft the manuscript. All authors read and approved the final manuscript.

ACKNOWLEDGEMENTS

We would like to thank Marcel Cardillo and Sebastian Duchene for the helpful discussions regarding the methods used in this work. This work was funded by the Australian Research Council.

REFERENCES

1. Huey RB: Rapid evolution of a geographic cline in size in an introduced fly. *Science* 2000, 287:308-309.

2. Rieseberg LH, Wendel J: Plant speciation - rise of the poor cousins. *New Phytol* 2003, 161:3-8.

3. Wu CI, Ting CT: Genes and speciation. *Nat Rev Genet* 2004, 5:114-122.

4. Allen Orr H: The genetics of species differences. *Trends Ecol Evol* 2001, 16:343-350.

5. Barraclough TG, Harvey PH, Nee S: Rate of rbcL gene sequence evolution and species diversification in flowering plants (Angiosperms). *Proc Roc Soc B Bio* 1996, 263:589-591.

6. Barraclough T, Savolainen V: Evolutionary rates and species diversity in flowering plants. *Evolution* 2001, 55:677-683.

7. Eo SH, DeWoody JA: Evolutionary rates of mitochondrial genomes correspond to diversification rates and to contemporary species richness in birds and reptiles. *Proc Roc Soc B Bio* 2010, 277:3587-3592.

8. Lanfear R, Ho SYW, Love D, Bromham L: Mutation rate is linked to diversification in birds. *Proc Natl Acad Sci* 2010, 107:20423-20428.

9. Fontanillas E, Welch JJ, Thomas JA, Bromham L: The influence of body size and net diversification rate on molecular evolution during the radiation of animal phyla. *BMC Evol Biol* 2007, 7:95.

10. Webster AJ, Payne RJH, Pagel M: Molecular phylogenies link rates of evolution and speciation. *Science* 2003, 301:478.

11. Goldie X, Lanfear R, Bromham L: Diversification and the rate of molecular evolution: no evidence of a link in mammals. *BMC Evol Biol* 2011, 11:286.

12. Rieseberg LH, Blackman BK: Speciation genes in plants. *Ann Bot* 2010, 106:439-455.

13. Pagel M, Venditti C, Meade A: Large punctuational contribution of speciation to evolutionary divergence at the molecular level. *Science* 2006, 314:119-121.

14. Venditti C, Pagel M: Speciation as an active force in promoting genetic evolution. *Trends Ecol Evol* 2010, 25:14-20.

15. Woolfit M, Bromham L: Increased rates of sequence evolution in endosymbiotic bacteria and fungi with small effective population sizes. *Mol Biol Evol* 2003, 20:1545-1555.

16. Orr H, Turelli M: The evolution of postzygotic isolation: accumulating Dobzhansky-Muller incompatibilities. *Evolution* 2001, 55:1085-1094.
17. Eo SH, Wares JP, Carroll JP: Population divergence in plant species reflects latitudinal biodiversity gradients. *Biol Letters* 2008, 4:382-384.
18. Schluter D, Conte GL: Genetics and ecological speciation. *Proc Natl Acad Sci* 2009, 106:9955-9962.
19. Spielman D, Brook BW, Frankham R: Most species are not driven to extinction before genetic factors impact them. *Proc Natl Acad Sci* 2004, 101:15261-15264.
20. Kisel Y, Moreno-Letelier A, Bogarín D: Testing the link between population genetic differentiation and clade diversification in Costa Rican orchids. *Evolution* 2012, 66:3035-3052.
21. Davies TJ, Savolainen V, Chase MW, Moat J, Barraclough TG: Environmental energy and evolutionary rates in flowering plants. *Proc Roc Soc B Bio* 2004, 271:2195-2200.
22. Verdú M: Age at maturity and diversification in woody angiosperms. *Evolution* 2002, 56:1352-1361.
23. Smith SA, Donoghue MJ: Rates of molecular evolution are linked to life history in flowering plants. *Science* 2008, 322:86-89.
24. Gaut B, Yang L, Takuno S, Eguiarte LE: The patterns and causes of variation in plant nucleotide substitution rates. *Annu Review Ecol Syst* 2011, 42:245-266.
25. Kimura M: *The Neutral Theory of Molecular Evolution*. Cambridge, UK: Cambridge University Press; 1985.
26. Greiner S, Rauwolf U, Meurer J, Herrmann RG: The role of plastids in plant speciation. *Mol Ecol* 2011, 20:671-691.
27. Kirk JTO, Tilney-Bassett RAE: *The plastids. Their chemistry, structure, growth and inheritance*. London & San Francisco: WH Freeman & Co.; 1967.
28. Stubbe W: Oenothera—An ideal system for studying the interactions of genome and plastome. *Plant Mol Biol Rep* 1989, 7:245-257.
29. Levin D: The cytoplasmic factor in plant speciation. *Syst Bot* 2003, 28:5-11.

30. Xiong AS, Peng RH, Zhuang J, Gao F, Zhu B, Fu XY, Xue Y, Jin XF, Tian YS, Zhao W, Yao QH: Gene duplication, transfer, and evolution in the chloroplast genome. *Biotechnol Adv* 2009, 27:340-347.
31. Wicke S, Schneeweiss GM, DePamphilis CW, Müller KF, Quandt D: The evolution of the plastid chromosome in land plants: gene content, gene order, gene function. *Plant Mol Biol* 2011, 76:273-297.
32. Herrmann RG, Maier RM, Schmitz-Linneweber C: Eukaryotic genome evolution: rearrangement and coevolution of compartmentalized genetic information. *Philos T Roy Soc B* 2003, 358:87-97.
33. Johnson NA: Hybrid incompatibility genes: remnants of a genomic battlefield? *Trends Genet* 2010, 26:317-325.
34. Lanfear R, Welch JJ, Bromham L: Watching the clock: studying variation in rates of molecular evolution between species. *Trends Ecol Evol* 2010, 25:495-503.
35. Cowling RM, Lamont BB: On the nature of Gondwanan species flocks: diversity of proteaceae in Mediterranean South-western Australia and South Africa. *Aust J Bot* 1998, 46:335.
36. Valente LM, Reeves G, Schnitzler J, Mason IP, Fay MF, Rebelo TG, Chase MW, Barraclough TG: Diversification of the African genus Protea (Proteaceae) in the Cape biodiversity hotspot and beyond: equal rates in different biomes. *Evolution* 2010, 64:745-760.
37. Prunier R, Holsinger KE: Was it an explosion? Using population genetics to explore the dynamics of a recent radiation within Protea (Proteaceae L.). *Mol Ecol* 2010, 19:3968-3980.
38. Sauquet H, Weston PH, Anderson CL, Barker NP, Cantrill DJ, Mast AR, Savolainen V:Contrasted patterns of hyperdiversification in Mediterranean hotspots. *Proc Natl Acad Sci* 2009, 106:221-225.
39. Lancaster LT: Molecular evolutionary rates predict both extinction and speciation in temperate angiosperm lineages. *BMC Evol Biol* 2010, 10:162.
40. Bromham L: Why do species vary in their rate of molecular evolution? *Biol Letters* 2009, 5:401-404.
41. Yang L, Gaut BS: Factors that contribute to variation in evolutionary rate among Arabidopsis genes. *Mol Biol Evol* 2011, 28:2359-2369.

42. Ikemura T: Codon usage and tRNA content in unicellular and multicellular organisms. *Mol Biol Evol* 1985, 2:13-34.
43. Sloan DB, Taylor DR: Testing for selection on synonymous sites in plant mitochondrial DNA: the role of codon bias and RNA editing. *J Mol Evol* 2010, 70:479-491.
44. Morton BR: Selection on the codon bias of chloroplast and cyanelle genes in different plant and algal lineages. *J Mol Evol* 1998, 46:449-459.
45. Fishman L, Willis JH: A cytonuclear incompatibility causes anther sterility in*Mimulus* hybrids. *Evolution* 2006, 60:1372-1381.
46. Rieseberg LH, Willis JH: Plant speciation. *Science* 2007, 317:910-914.
47. Welch J: Accumulating Dobzhansky-Muller incompatibilities: reconciling theory and data. *Evolution* 2004, 58:1145-1156.
48. Eyre-Walker A, Gaut BS: Correlated rates of synonymous site evolution across plant genomes. *Mol Biol Evol* 1997, 14:455-460.
49. Wright S, Keeling J, Gillman L: The road from Santa Rosalia: a faster tempo of evolution in tropical climates. *Proc Natl Acad Sci* 2006, 103:7718-7722.
50. Allen AP, Brown JH, Gillooly JF: Global biodiversity, biochemical kinetics, and the energetic-equivalence rule. *Science* 2002, 297:1545-1548.
51. Rohde K: Latitudinal gradients in species diversity: the search for the primary cause. *Oikos* 1992, 65:514-527.
52. Charlesworth B: Fundamental concepts in genetics: effective population size and patterns of molecular evolution and variation. *Nat Rev Genet* 2009, 10:195-205.
53. Lee CE: Evolutionary genetics of invasive species. *Trends Ecol Evol* 2002, 17:386-391.
54. Carson HL, Templeton AR: Genetic Revolutions in Relation to Speciation Phenomena: The Founding of New Populations. *Annu Rev Ecol Syst* 1984, 15:97-132.
55. Ohta T: The nearly neutral theory of molecular evolution. *Annu Rev Ecol Syst* 1992, 23:263-286.
56. Woolfit M: Effective population size and the rate and pattern of nucleotide substitutions. *Biol letters* 2009, 5:417-420.

57. Eyre-Walker A: The genomic rate of adaptive evolution. *Trends Ecol Evol* 2006, 21:569-575.

58. Charlesworth J, Eyre-Walker A: The other side of the nearly neutral theory, evidence of slightly advantageous back-mutations. *Proc Natl Acad Sci* 2007, 104:16992-16997.

59. Servedio MR, Van Doorn GS, Kopp M, Frame AM, Nosil P: Magic traits in speciation: "magic" but not rare? *Trends Ecol Evol* 2011, 26:389-397.

60. Nosil P, Schluter D: The genes underlying the process of speciation. *Trends Ecol Evol* 2011, 26:160-167.

61. Strasburg JL, Sherman NA, Wright KM, Moyle LC, Willis JH, Rieseberg LH: What can patterns of differentiation across plant genomes tell us about adaptation and speciation? *Philos T Roy Soc B* 2012, 367:364-373.

62. Kapralov MV, Filatov DA: Widespread positive selection in the photosynthetic Rubisco enzyme. *BMC Evol Biol* 2007, 7:73.

63. Hao DC, Chen SL, Xiao PG: Molecular evolution and positive Darwinian selection of the chloroplast maturase matK. *J Plant Res* 2010, 123:241-247.

64. Phillips R, Dixon K, Peakall R: Low population genetic differentiation in the Orchidaceae: implications for the diversification of the family. *Mol Ecol* 2012, 21:5208-5220.

65. Darwin C: *On the Origins of Species by Means of Natural Selection*. London: John Murray; 1859.

66. Presgraves DC: Speciation genetics: search for the missing snowball. *Current Biol* 2010, 20:1073-1074.

67. Gavrilets S: Perspective: Models of speciation: What have we learned in 40 years? *Evolution* 2003, 57:2197-2215.

68. Gourbière S, Mallet J: Are species real? The shape of the species boundary with exponential failure, reinforcement, and the "missing snowball". *Evolution* 2010, 64:1-24.

69. Matute DR, Butler IA, Turissini DA, Coyne JA: A test of the snowball theory for the rate of evolution of hybrid incompatibilities. *Science* 2010, 329:1518-1521.

70. Moison M, Roux F, Quadrado M, Duval R, Ekovich M, Lê D-H, Verzaux M, Budar F:Cytoplasmic phylogeny and evidence of

cyto-nuclear co-adaptation in Arabidopsis thaliana. *Plant J* 2010, 63:728-738.

71. Abbott R, Albach D, Ansell S, Arntzen JW, Baird SJE, Bierne N, Boughman J, Brelsford A, Buerkle CA, Buggs R, Butlin RK, Dieckmann U, Eroukhmanoff F, Grill A, Cahan SH, Hermansen JS, Hewitt G, Hudson AG, Jiggins C, Jones J, Keller B, Marczewski T, Mallet J, Martinez-Rodriguez P, Möst M, Mullen S, Nichols R, Nolte AW, Parisod C, Pfennig K: Hybridization and speciation. *J Evol Bio* 2013, 26:229-246.
72. Ainouche ML, Jenczewski E: Focus on polyploidy. *New Phytol* 2010, 186:1-4.
73. Petit RJ, Duminil J, Fineschi S, Hampe A, Salvini D, Vendramin GG: Comparative organization of chloroplast, mitochondrial and nuclear diversity in plant populations. *Mol Ecol* 2005, 14:689-701.
74. Maréchal A, Brisson N: Recombination and the maintenance of plant organelle genome stability. *New Phytol* 2010, 186:299-317.
75. Bock R: The give-and-take of DNA: horizontal gene transfer in plants. *Trends Plant Sci* 2010, 15:11-22.
76. Mallet J: Hybrid speciation. *Nature* 2007, 446:279-283.
77. Soltis DE, Albert VA, Leebens-Mack J, Bell CD, Paterson AH, Zheng C, Sankoff D, Depamphilis CW, Wall PK, Soltis PS: Polyploidy and angiosperm diversification. *Am J Bot* 2009, 96:336-348.
78. Weston P, Barker N: A new suprageneric classification of the Proteaceae, with an annotated checklist of genera. *Telopea* 2006, 11:314-344.
79. Hugall AF, Lee MSY: The likelihood node density effect and consequences for evolutionary studies of molecular rates. *Evolution* 2007, 61:2293-2307.
80. Gouy M, Guindon S, Gascuel O: SeaView version 4: A multiplatform graphical user interface for sequence alignment and phylogenetic tree building. *Mol Biol Evol* 2010, 27:221-224.
81. Paradis E, Claude J, Strimmer K: APE: Analyses of Phylogenetics and Evolution in R language. *Bioinformatics* 2004, 20:289-290.

82. Zwickl DJ: *GARLI: genetic algorithm for rapid likelihood inference*. 2006. https://www.nescent.org/wg_garli/

83. Pond SLK, Frost SDW, Muse SV: HyPhy: hypothesis testing using phylogenies. *Bioinformatics* 2005, 21:676-679.

84. Muse S, Gaut B: A likelihood approach for comparing synonymous and nonsynonymous nucleotide substitution rates, with application to the chloroplast genome. *Mol Biol Evol* 1994, 11:715-724.

85. Wilcoxon F, Katti SK, Wilcox RA: *Critical Values and Probability Levels for the Wilcoxon Rank Sum Test and the Wilcoxon Signed Rank Test*. American Cyanamid: Lederle Laboratories Division; 1963.

Chapter 7

A Survey of Nuclear Ribosomal Internal Transcribed Spacer Substitution Rates Across Angiosperms: An Approximate Molecular Clock with Life History Effects

Kathleen M Kay[1, 2], Justen B Whittall[2, 3], and Scott A Hodges[2]

[1]Department of Plant Biology, Michigan State University, 166 Plant Biology Building, East Lansing, MI, 48824, USA

[2]Department of Ecology, Evolution and Marine Biology, University of California, Santa Barbara, CA, 93106, USA

[3]Section of Evolution and Ecology, University of California, One Shields Ave., Davis, CA, 95616, USA

ABSTRACT

Background

A full understanding of the patterns and processes of biological diversification requires the dating of evolutionary events, yet the fossil record is inadequate for most lineages under study. Alternatively, a molecular clock approach, in which DNA or amino acid substitution rates are calibrated with fossils or geological/climatic events, can provide indirect estimates of clade ages and diversification rates. The utility of this approach depends on the rate constancy of molecular evolution at a genetic locus across time and across lineages. Although the nuclear ribosomal internal transcribed spacer region (nrITS) is increasingly being used to infer clade ages in plants, little is known about the sources or magnitude of variation in its substitution rate. Here, we systematically review the literature to assess substitution rate variation in nrITS among angiosperms, and we evaluate possible correlates of the variation.

Results

We summarize 28 independently calibrated nrITS substitution rates ranging from 0.38×10^{-9} to 8.34×10^{-9} substitutions/site/yr. We find that herbaceous lineages have substitution rates almost twice as high as woody plants, on average. We do not find any among-lineage phylogenetic constraint to the rates, or any effect of the type of calibration used. Within life history categories, both the magnitude of the rates and the variance among rates tend to decrease with calibration age.

Conclusion

Angiosperm nrITS substitution rates vary by approximately an order of magnitude, and some of this variation can be attributed to life history categories. We make cautious recommendations for the use of nrITS as an approximate plant molecular clock, including an outline of more appropriate phylogenetic methodology and caveats against over interpretation of results. We also suggest that for lineages with independent calibrations, much of the variation in nrITS substitution rates may come from uncertainty in calibration date estimates, highlighting the importance of accurate and/or multiple calibration dates.

BACKGROUND

Decades after the idea of a molecular clock was first broached[1], the weight of evidence suggests that rate heterogeneity, not rate constancy, is the norm in molecular evolution (reviewed in [2,3]) and that an exact molecular clock will not be found. Nevertheless, an approximate clock can be established for a locus with many independent calibrations [4], and would be useful for roughly dating lineages and testing time-dependent evolutionary hypotheses. Local rate-constancy and even a single calibration can allow dating within a particular lineage, but much of the excitement surrounding molecular clocks stems from the possibility of approximating dates for the majority of lineages that lack independent calibrations. This has become commonplace in the animal kingdom, especially with the widely applied mtDNA clock [5,6], but a general molecular clock able to date relatively recent divergences in plants remains elusive. Because there will be variation among independently calibrated substitution rates, a first step in determining whether a particular locus may be useful as a molecular clock is to ascertain the amount and sources of this rate variation.

In the last decade, the nuclear ribosomal internal transcribed spacer region (nrITS) has revolutionized species-level plant phylogenetics. Because concerted evolution has generally homogenized sequence variation among the numerous ribosomal DNA copies within an individual, direct sequencing of this region is possible for most systems. This, coupled with the availability of universal primers and elevated substitution rates compared to most chloroplast regions, make it especially accessible and appropriate for resolving interspecific phylogenetic relationships [7,8]. Although reliance on nrITS as the sole source of phylogenetic evidence has come under criticism because of certain features of its evolution [9], it remains the most efficient locus for generating species-level phylogenetic inferences in most plant groups.

The widespread use of the nrITS region at the species level makes it a good candidate for a potential plant molecular clock. Substitution rates have been independently calibrated for a number of plant groups with diverse life histories and growth forms that represent distinct branches of the angiosperm phylogeny. Preliminary surveys of some of these rates suggest that they vary by almost an order of magnitude [10,11]. However, these limited surveys do not fully evaluate the variation in available nrITS substitution rates nor do they examine possible sources of this variation. Even without a full summary of nrITS rate variation, there is a growing trend in the literature to utilize some published rates for groups that lack their own independent calibration, either by choosing a rate based on life history similarity or phylogenetic relatedness, or by using a range of the more commonly-cited rates, such as Suh (1993) or Sang (1994, 1995). Here, we conduct a systematic search of the literature in order to identify and evaluate more thoroughly nrITS rate variation. Specifically, we report a summary of independently calibrated rates, examine potential correlates that may account for some of the variation in rates, and make conservative recommendations regarding the use of nrITS as an approximate molecular clock for angiosperms.

RESULTS

Literature Survey

We identified 29 independent nrITS substitution rates ranging from 0.38 $\times 10^{-9}$ subs/site/yr in *Hamamelis* to 19 $\times 10^{-9}$ subs/site/yr in *Gentiana* Sect. *Ciminalis* (Fig. 1; Table 1). The latter rate was excluded as an outlier from all of our analyses since the calibration was based on only a single base-pair substitution and the rate was greater than double the second highest rate of 8.34 $\times 10^{-9}$subs/site/year in *Soldanella*. The mean of the 28 remaining rates was 2.86 $\times 10^{-9}$ subs/site/yr. Of those studies reporting results of rate constancy tests, 12 out of 19 passed or partially passed (Table 1).

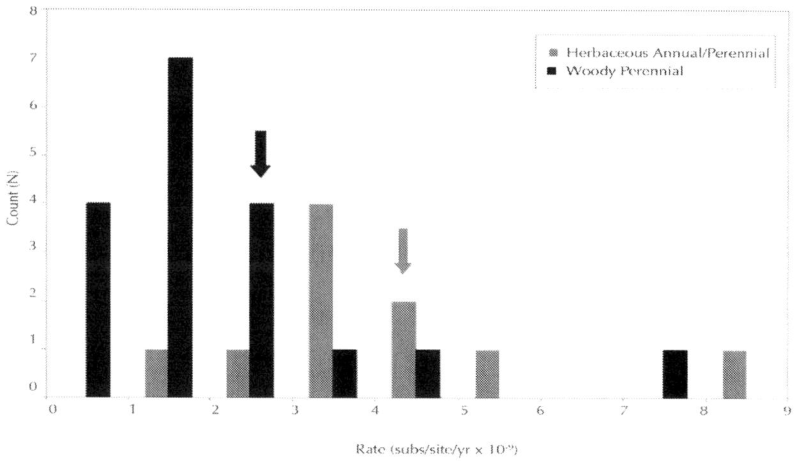

Figure 1: The distribution of ITS rates. A frequency histogram of nrITS rates, showing the difference between woody perennial and herbaceous annual/perennial rates. Arrows indicate the average rates for each life history category. The outlier rate for *Gentiana* Sect. *Cimialis* (19×10^{-9} subs/site/yr) has been removed.

Table 1: Independently calibrated nrITS substitution rates sorted by magnitude of rate

Taxon	Family	Life History[a]	Clock Test	Calibration Type[b]	Calibration Date (ma)	Rate[d]	Reference
Hamamelis	Hamamelidaceae	W	NA	G	8.5	0.38	[48]
Winteraceae	Winteraceae	W	NA	G	65[c]	0.45[e]	[27]
Nothofagus	Nothofagaceae	W	NA	F	83	0.50	[49]
Salicaceae	Salicaceae	W	NA	F	50[c]	0.60	[50]
Aralia sect. Dimorphanthus	Araliaceae	W	NA	G	12	1.07	[51]
Echium	Boraginaceae	W	Passed	G	20	1.10	[52]
Alnus	Betulaceae	W	Rejected	F	70	1.10	[53]
Empetraceae	Empetraceae	W	Passed	F	37	1.44	[54]
Saxifraga	Saxifragaceae	H	NA	NA	5.5[c]	1.72	[55]
Aesculus	Hippocastanaceae	W	Passed	F	65	1.72	[56]
Gaertnera	Rubiaceae	W	Passed	F	54	1.99	[11]
Ormocarpum	Fabaceae	W	Rejected	G	35	2.00	[57]
Inga	Fabaceae	W	Rejected	G	3.5	2.34	[10]
Phylica	Rhamnaceae	W	Rejected	G	2	2.44	[58]
Adansonia	Bombacaceae	W	Passed	F	47[c]	2.48	[59]
Eupatorium	Asteraceae	H	Passed	M (ndhF)	14.8	2.51	[60]
Tarweeds/ Hawaiian silverswords	Asteraceae	W	Passed	C	15	3.00	[61]

Robinioid legumes	Fabaceae	W	Rejected	F	39.4c	3.30	[62]
Lupinus	Fabaceae	H	NA	F	60	3.46e	[63]
Astragalus	Fabaceae	H	NA	F	35	3.50	[64]
Cucurbitoideae	Cucurbitaceae	H	NA	F	40	3.62	[65]
Ehrharta	Poaceae	H	Rejected	M (rbcL) & F	41	3.81	[66]
Plantago	Plantaginaceae	H	Rejected	G	0.6	4.27	[71]
Gentianella	Gentianaceae	H	Passed	G & F	3	4.52	[67]
Dendroseris	Asteraceae	W	Passed	G & M (cpDNA)	3.3c	5.00	[28]
Gossypium	Malvaceae	H	NA	F & M (cpDNA)	8.5c	5.50e	[68]
Robinsonia	Asteraceae	W	Passed	G	4	7.83	[69]
Soldanella	Primulaceae	H	Passed	F	23.3	8.34	[70]
Gentiana sect. Ciminalis	Gentianaceae	H	Passed	G	0.1	19.00	[25]

[a]W = woody, H = herbaceous.

[b]G = geographic vicariance; F = fossil; M = molecular clock from a different locus, followed by an identifier for that locus in parentheses; C = climatic.

[c]Calibration age and associated substitution rate are averages of two reported ages.

[d]substitutions/site/year × 10-9.

[e]Substitution rate is a weighted average of ITS1 and ITS2 rates reported separately.

Kay et al. BMC Evolutionary Biology 2006 6:36, doi:10.1186/1471-2148-6-36

Phylogenetic Signal

Twenty-one different angiosperm families are represented among the 29 rates, which comprise non-overlapping portions of the Angiosperm evolutionary tree. No significant phylogenetic signal to the rates was detected ($p = 0.393$), suggesting that phylogenetic relatedness is not an appropriate justification when choosing rates from the literature. Indeed, rates from the family Asteraceae alone span almost the entire range found in Table 1, from 2.5×10^{-9} subs/site/year in *Eupatorium* to 7.83×10^{-9} subs/site/year in *Robinsonia*.

Life History Effects

The average nrITS substitution rates for herbaceous and woody lineages are significantly different ($p = 0.0013$, Wilcoxon Rank Sum Test; Fig. 1). The herbaceous annual/perennial rates ranged from 1.72×10^{-9} to 8.34×10^{-9} substitutions/site/year (mean = 4.13×10^{-9} subs/site/yr, $N = 10$), and the woody perennial rates ranged from 0.38×10^{-9} to 7.83×10^{-9} substitutions/site/year (mean = 2.15×10^{-9} subs/site/yr, $N = 18$). In our analysis of phylogenetic independent contrasts, we found that nine of ten life history contrasts exhibit a lower rate associated with the woody perennial lineage (Sign Test, $p < 0.01$). The single opposing contrast involved the Hawaiian silverswords. This rate is based on a clade that includes both the woody perennial silverswords and the herbaceous perennial tarweeds. Although initially coded as woody perennials, this state is considered the derived condition among the tarweeds. Therefore, the rate may be more representative of the herbaceous perennial ancestry in the clade than the derived woody perennial silversword lineage. Our contrast results are robust to either life history classification of the tarweed/silversword rate (alternative coding, $p < 0.01$).

DISCUSSION

Range of nrITS Substitution Rates

This study represents the most comprehensive collection of nrITS substitution rates to date. The largest previous survey [10] included independently calibrated rates from 12 angiosperm lineages. They recommended using a range of rates from 1.72×10^{-9} to 7.83×10^{-9} for dating the diversification of *Inga*, after removing the rate from Winteraceae because of its longer generation time and possible error associated with the calibration date. The addition of 16 independently calibrated rates in our study introduces seven new rates slower than 1.72×10^{-9}, one of which is slower than the Winteraceae rate. Our broader sampling of nrITS substitution rates allowed us to analyze more fully potential sources of variation in these rates.

Life History Effects

Striking differences in rates of nrITS evolution between life history categories, with annuals/herbaceous perennials exhibiting significantly higher rates, are consistent with a general pattern of higher substitution rates for annual compared to perennial plants and for herbs compared to woody trees that has been reported from chloroplast [12-16] and mitochondrial studies [17]. Life history has also been suggested to affect rates specifically of nrITS substitution ([18,19] but see [20]), with annual species exhibiting longer nrITS branch lengths than perennials. In a more comprehensive review, however, Whittle and Johnston [21] found no difference in nrITS branch lengths between sister taxa differing in annual versus perennial life history. Our study may be more likely to detect rate differences associated with life history because we are using more distinctly polarized life history categories and our comparisons are across deeper phylogenetic splits, with more time for rate differences to accumulate.

Although the underlying mechanism is unclear, our results agree with previous studies in which shorter generation times are associated with faster rates of molecular evolution [22,23]. It is difficult to explain how generation time would affect substitution rates in plants, which do not sequester their germ lines and therefore can pass on somatic mutations to their gametes, unless the majority of inherited mutations arise during meiosis and not mitosis. Furthermore, we do not know whether generation time per se is a causative factor of substitution rate variation. Differences in generation time in plants are confounded by correlated differences in growth form [21] and mating system [24], such that perennials are the only plants forming extensive wood and annuals are much more frequently self-fertilizing. A mechanistic understanding of the effects of these factors on rates of molecular evolution will increase the power of molecular clock approaches in plant evolution.

Errors in Rate Estimation

A potentially confounding factor associated with our life history classes is the type of calibration point used and its potential associated rate biases. Fossil data typically provide a minimum age of a particular lineage, and therefore may be more likely to *overestimate* a substitution rate, while biogeographic or climatic events provide a maximum age and may be more likely to*underestimate* a rate. Thus, an unacknowledged correlation between life history and calibration type potentially could confound our results for rate differences between woody and herbaceous plants. These potential biases appear not to be an issue. First, our dataset is well balanced with fossil calibrated rates for eight woody perennial and for six herbaceous annual/perennial groups. There is no overall correlation between type of calibration and life history category (^{2} = 0.156, ns). Second, within each life history category, we see no apparent bias due to the type of calibration. For example, within the herbaceous rates, only one of the five highest rates is calibrated exclusively with fossil evidence. Similarly, of the five slowest woody perennial rates, three are biogeographic and two are fossil calibrated.

The age of the calibration point may affect the precision of the substitution rate estimates, since more recent calibration ages are based on fewer "ticks" of the molecular clock. Older lineages will have more nucleotide substitutions and thus these rate estimates should become more precise, at least until they become biased by sequence saturation, or multiple substitutions at the same site. Among our 29 original rates, calibration ages ranged from 0.1 Ma to 83 Ma. There appears to be substantially more variation in substitution rates based on recent calibration points compared to those from older calibration ages in both life history categories (Fig. 2). An extreme example is the exceptionally high rate from *Gentiana* sect. *Ciminalis*, that was more than twice as high as any other rate. This rate was calibrated based on a biogeographic split of only 0.1 Ma after which only one nucleotide substitution occurred [25]. When rate is regressed on age of calibration, separately for each life history category, the residuals exhibit a negative trend with increasing calibration age indicating lower variance for older calibration ages (Fig. 2). Another indication of this trend is found by comparing the variation in residuals for rates with calibration points older and younger than 30 Ma. The mean absolute value of these residuals in the more recently calibrated half is more than three times higher than in the half with older calibrations.

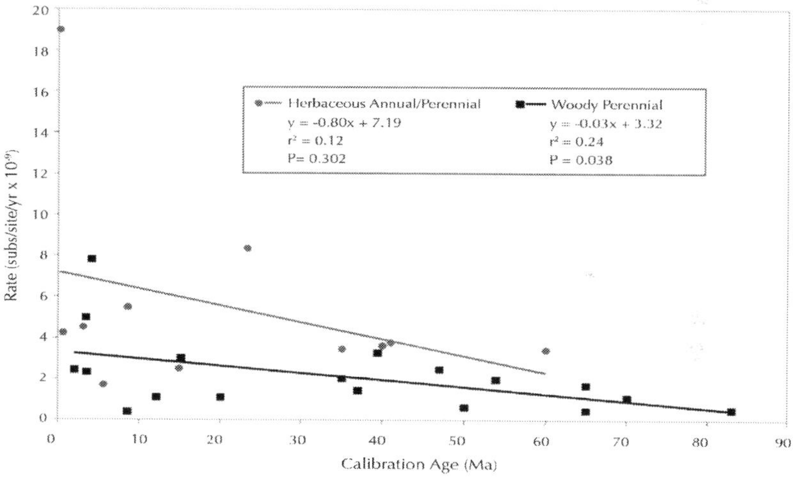

Figure 2: The effects of calibration age. A comparison of nrITS substitution rates with calibration age reveals higher variance at younger calibration points. The decreasing substitution rate with older calibration ages is only significant for the woody perennials. The herbaceous annual/perennial regression is not significant and becomes even less so when the outlier is removed ($p = 0.861$).

We also found a trend of decreasing rate with the age of calibration, which was significant among woody perennials (Fig. 2). This trend may be due to the models of sequence evolution utilized for estimating sequence divergence. Without correcting for sequence saturation, rate estimates are expected to decrease with increasing time since divergence [5,6]. In molecular clock calculations, using a model of sequence evolution that does not fully capture sequence saturation and among-site rate variation potentially can lead to large errors in divergence time estimates (reviewed in [3]). Most of the original rates we found used the relatively simple Jukes-Cantor or Kimura 2-parameter models, so this might be a significant issue, particularly for studies with older calibration dates. Ideally, objective criteria for model selection, such as the likelihood ratio tests easily implemented in MODELTEST [26] should be used before calculating a rate.

As a post hoc exploration of the effects of methods of estimating sequence divergence, including the choice of substitution model and the treatment of rate heterogeneity among branches, we reanalyzed the sequence data used to estimate two of the nrITS rates, for the Winteraceae [27] with one of the oldest calibration dates and for *Dendroseris* [28] with a recent calibration. These are also two of the most frequently cited rates in the literature. For the Winteraceae, our resulting substitution rate estimates, based on a TrN+G substitution model and penalized likelihood rate smoothing, are higher than crude estimates based on uncorrected counts of pairwise differences. However, the uncertainty reported for their biogeographic calibration – the isolation of Pseudowinteraceae in Australia – for which estimates range from 50 to 80 Ma, affected the overall rate estimates substantially more than the different methods of estimating branch lengths (substitutions/site/year × 10^{-9}: 50 Ma calibration crude = 0.52, corrected = 0.65; 80 Ma calibration crude = 0.33, corrected = 0.41). Importantly, for this group with one of the oldest calibrations, which should be more sensitive to the effects of sequence saturation and rate heterogeneity, branch length estimation methods make a small difference in light of the overall range of nrITS substitution rates in our survey, not even affecting its category in the histogram of Figure 1. For *Dendroseris*, there was no significant rate heterogeneity, and using a more complex substitution model had no effect on the substitution rate estimate. In contrast, there was substantial variation in rates due to uncertainty in the biogeographic calibration, ranging from 4 Ma based on the age of Masatierra Island to 2.6 Ma based on a chloroplast DNA clock calculation (substitutions/site/year × 10^{-9}: 4 Ma calibration = 3.94; 2.6 Ma calibration = 6.06). These reanalyses suggest that fine-tuning the estimates of molecular substitution makes a relatively small change to the overall rate estimate in light of substantial uncertainty in calibration dates. While branch length estimates clearly should be optimized, there is also a real need for more accurate and multiple calibrations to reduce the uncertainty of rate estimates [29].

An Approximate Plant Molecular Clock?

The variance we find in nrITS substitution rates is likely the result of both error in individual rate estimates and biological differences among lineages in actual substitution rates. As noted above, substantial error in estimating individual rates can come from uncertainty in the dating of the fossil or biogeographic/climatic event used as a calibration and in the assignment of that calibration to a node in the phylogeny [29,30]. Further error is contributed by uncertainty in the phylogeny itself, both in the branching topology and the branch length estimates [31]. Another issue is that gene trees are not identical to species trees [3]. Coalescence of nrITS alleles should predate species divergences by a factor related to the effective population size, and any reticulate evolution or lineage sorting could obscure the true tree topology [32]. In attempting to date a particular plant lineage, few studies adequately consider these sources of error.

While some of the variance in the rates we found likely comes from these sources of error, there are also very real differences in nrITS substitution rates within and among lineages. Statistically significant rate heterogeneity was found in seven of the nineteen studies reporting tests of rate constancy. Furthermore, in spite of all the sources of noise in our dataset, we found a substantial effect of life history, with long-lived woody plants exhibiting much slower substitution rates than shorter-lived herbaceous lineages. Thus even if error in substitution rate estimates could be eliminated, the existence of a universal nrITS substitution rate is highly unlikely.

Despite the many sources of error in rate estimates, however, we find a limited distribution of rates with a central tendency, especially when separated by life history category (Fig. 1). We therefore support the cautious use of this locus as a general molecular clock for roughly dating events in plant evolution, but we offer a few important caveats. Borrowing just one or a few rates from the literature, especially based on phylogenetic relatedness, does not adequately account for the nature and extent of variation in the nrITS "clock". Instead, to roughly estimate a date we suggest that, within a life history category, the entire range of rates be considered. Furthermore, before using any rate for dating historical events, whether calculated from an internal calibration or borrowed from the literature, nrITS rate constancy should be tested within the group of interest, ideally with a tree-based likelihood ratio test [33,34], and any lack of local rate constancy should be considered in the dating methodology [35].

The distribution of rates could then be used to examine specific hypotheses for the timing of events for lineages without calibrations of their own. For example, if there is a hypothesis for the age of a particular node in the phylogeny of a group of taxa with an nrITS dataset, one could calculate the required substitution rate to fit that age to the node and then compare the required rate to the distribution of rates presented here. A good example of this type of hypothesis testing is found in the genus *Aquilegia* (Ranunculaceae). Hummingbird pollination in this group was suggested to have originated during the mid-Pliocene (ca. 3.5 Ma) based on floristic associations [36]. Hodges et al. [37] calculated the nrITS substitution rate necessary within *Aquilegia* to fit the hypothesized age of the hummingbird pollination syndrome (0.11% per my = 0.56×10^{-9}subs/site/yr). This is approximately ten times slower than the average annual/perennial herbaceous rate (4.13×10^{-9}) and more than three times slower than the slowest annual/perennial herbaceous value. Even using the slowest annual/perennial rate (1.72×10^{-9}), the estimated age of hummingbird pollination is only half as old as predicted (approximately 1.66 Ma). Without any justification for such an unusually slow nrITS substitution rate in *Aquilegia*, hummingbird pollination likely has evolved more recently than originally predicted [37].

CONCLUSIONS

The results from our literature survey underscore the popularity of the nrITS region as a tool to date relatively recent historical events in plant evolution. The original rates that we found came from a broad taxonomic spectrum representing a diversity of life histories and were calculated using a variety of calibrations and methodologies.

Nevertheless, the rates were consistent to approximately an order of magnitude, or even less for herbaceous lineages considered separately. Although this survey should be considered an ongoing project, given these initial results, the nrITS may be useful as an approximate molecular clock for roughly dating divergences, calculating diversification rates, and hypothesis testing in plant groups that currently lack a fossil or biogeographic calibration of their own.

METHODS

Literature Survey

We surveyed *Systematic Botany, Molecular Phylogenetics and Evolution, The American Journal of Botany, Evolution* and any references therein, from January 1995 through December 2003. We recorded rates for any study that independently calibrated and reported a rate for ITS1 and ITS2. If multiple rates were reported, we followed the author's recommendation on which was likely to be most accurate, unless their reasoning was based on the rate's similarity to other published rates. In such cases, we took the average of the reported rates. When separate rates were reported for ITS1 and ITS2, a weighted average based on the average length of each nrITS region was calculated. For each rate, we also recorded the predominant growth form/life history [38,39], rate constancy test results, calibration type, and calibration age.

Phylogenetic Signal

Frequently, authors that utilize published rates in their own studies base their choices on phylogenetic relatedness. In order to determine whether such choices are warranted, we tested for any phylogenetic signal in the rates. First we constructed a phylogeny of the taxa represented in our rates survey using the web-based program Phylomatic [40] and then manually edited the tree to reflect recent and more detailed phylogenetic hypotheses for some families [41-43] (tree available upon request). The Test for Serial Independence was implemented in the program Phylogenetic Independence v. 2.0 [44,45] with 1000 randomizations. This test compares the autocorrelation of adjacent branches of the original tree to that of a series of randomized trees.

Life History Effects

Generation time has been suggested to affect rates of molecular evolution [22,23], and thus we classified each rate according to the predominant life history of the clade, either annual/herbaceous perennial or woody perennial, as a proxy for generation time. The herbaceous category consists of relatively short-lived plants that reach sexual maturity within one to a few years, while the woody perennial life history category consists of long-lived lineages. It is possible that phylogenetic relatedness and life history could be confounded. Thus, to be conservative we also tested for rate differences due to life history using phylogenetically independent contrasts. Contrasts were identified with the BRUNCH algorithm in CAIC [46], using the same cladogram as in the above mentioned Test for Serial Independence. BRUNCH locates nodes for which there is a contrast in life history, and then calculates the direction of the rate difference across that node, using each taxon in no more than one contrast. It assumes that if sister taxa have the same life history state their most recent common ancestor also had that state, and that, when branch lengths are unknown, the rate for this ancestor can be calculated as the average rate of the derived taxa.

Reanalysis of Winteraceae and *Dendroseris* Examples

We reanalyzed the sequence data used to estimate the nrITS rates for the Winteraceae (GenBank accessions AY004111-AY004128; [27] Additional File 1) and for *Dendroseris* [28] Additional File 2. We aligned the sequences with ClustalX and refined the alignments manually. Phylogenetic relationships were estimated in PAUP 4.0b10 [47] using maximum likelihood and a TrN+G substitution model for Winteraceae and a TrNef model for *Dendroseris*, as chosen by MODELTEST. For Winteraceae, we found significant substitution rate heterogeneity in nrITS using a tree-wide likelihood ratio test for differences between clock-constrained and clock-unconstrained trees ($p < 0.01$; [34]). We then used penalized likelihood rate smoothing implemented in R8S [35] to estimate branch lengths and substitution rates in the face of this heterogeneity. Within*Dendroseris*, we did not find significant rate heterogeneity with a tree-wide LRT ($p > 0.1$), and therefore rate smoothing was unnecessary for estimating substitution rates.

AUTHORS' CONTRIBUTIONS

JW and KK designed and coordinated the study, carried out the literature survey, performed the analyses, and drafted the manuscript. SH provided original direction and helped to revise the manuscript. All authors discussed and interpreted the results and read and approved the final manuscript.

ACKNOWLEDGEMENTS

We thank D. Morrow for advice on analyses, J. Yang, T. Near and M. Sanderson for helpful discussions, and B. Igic for helpful comments and criticisms on an earlier version of this manuscript.

REFERENCES

1. Zuckerkandl E, Pauling L: Evolutionary divergence and convergence in proteins. In*Evolving Genes and Proteins*. Edited by Bryson V, Vogel HJ. New York: Academic Press; 1965:97-166.
2. Sanderson MJ: Estimating rate and time in molecular phylogenies: Beyond the molecular clock? In *Molecular Systematics of Plants II: DNA Sequencing*. Edited by Soltis DE, Soltis PS, Doyle JJ. Boston: Kluwer Academic Publishers; 1998:242-264.
3. Arbogast BS, Edwards SV, Wakeley J, Beerli P, Slowinski JB: Estimating divergence times from molecular data on phylogenetic and population genetic timescales. *Annual Review of Ecology and Systematics* 2002, 33:707-740.
4. Ayala F: Molecular clock mirages. *Bioessays* 1999, 21:71-75.
5. Brown WM, George M, Wilson AC: Rapid evolution of animal mitochondrial DNA. *Proceedings of the National Academy of Sciences of the United States of America* 1979,76(4):1967-1971.
6. Brown WM, Prager EM, Wang A, Wilson AC: Mitochondrial DNA sequences of primates: tempo and mode of evolution. *Journal of*

Molecular Evolution 1982, 18:225-239.

7. Small RL, Ryburn JA, Cronn RC, Seelanan T, Wendel JF: The tortoise and the hare: Choosing between noncoding plastome and nuclear Adh sequences for phylogeny reconstruction in a recently diverged plant group. *American Journal of Botany* 1998, 85(9):1301-1315.
8. Baldwin BG, Sanderson MJ, Porter JM, Wojciechowski MF, Campbell CS, Donoghue MJ: The ITS region of nuclear ribosomal DNA: A valuable source of evidence on angiosperm phylogeny. *Annals of the Missouri Botanical Garden* 1995, 82(2):247-277.
9. Alvarez I, Wendel JF: Ribosomal ITS sequences and plant phylogenetic inference. *Molecular Phylogenetics and Evolution* 2003, 29(3):417-434.
10. Richardson JE, Pennington RT, Pennington TD, Hollingsworth PM: Rapid diversification of a species-rich genus of Neotropical rain forest trees. *Science* 2001, 293(5538):2242-2245.
11. Malcomber ST: Phylogeny of *Gaertnera* Lam. (Rubiaceae) based on multiple DNA markers: evidence of a rapid radiation in a widespread, morphologically diverse genus. *Evolution* 2002, 56(1):42-57.
12. Brunsfeld SJ, Soltis PS, Soltis DE, Gadek PA, Quinn CJ, Strenge DD, Ranker TA:Phylogenetic relationships among the genera of Taxodiaceae and Cupressaceae: evidence from rbcL sequences. *Systematic Botany* 1994, 19(2):253-262.
13. Conti E, Fischbach A, Sytsma KJ: Tribal relationships in Onagraceae: implications from rbcL sequence data. *Annals of the Missouri Botanical Garden* 1993, 80(3):672-685.
14. Bousquet J, Strauss SH, Doerksen AH, Price RA: Extensive variation in evolutionary rate of rbcL gene sequences among seed plants. *Proc Natl Acad Sci U S A* 1992, 89(16):7844-7848.
15. Gaut BS, Muse SV, Clark WD, Clegg MT: Relative rates of nucleotide substitution at the*rbcL* locus of monocotyledonous plants. *Journal of Molecular Evolution* 1992, 35:292-303.
16. Wilson M, Gaut B, Clegg M: Chloroplast DNA evolves slowly in the palm family (Arecaceae). *Mol Biol Evol* 1990, 7(4):303-314.
17. Laroche J, Li P, Maggia L, Bousquet J: Molecular evolution of

angiosperm mitochondrial introns and exons. *Proceedings of the National Academy of Sciences of the United States of America* 1997,94(11):5722-5727.

18. Andreasen K, Baldwin BG: Unequal evolutionary rates between annual and perennial lineages of checker mallows (*Sidalcea*, Malvaceae): Evidence from 18S-26S rDNA internal and external transcribed spacers. *Mol Biol Evol* 2001, 18(6):936-944.
19. Ainouche AK, Bayer RJ: Phylogenetic relationships in *Lupinus* (Fabaceae : Papilionoideae) based on internal transcribed spacer sequences (ITS) of nuclear ribosomal DNA. *American Journal of Botany* 1999, 86(4):590-607.
20. Andreasen K: Implications of molecular systematic analyses on the conservation of rare and threatened taxa: Contrasting examples from Malvaceae. *Conservation Genetics* 2005, 6(3):399-412.
21. Whittle C-A, Johnston MO: Broad-scale analysis contradicts the theory that generation time affects molecular evolutionary rates in plants. *Journal of Molecular Evolution* 2003, 56(2):223-233.
22. Gu X, Li W-H: Higher rates of amino acid substitution in rodents than in humans. *Molecular Phylogenetics and Evolution* 1992, 1:211-214.
23. Ohta T: An examination of generation time effect on molecular evolution. *Proc Natl Acad Sci U S A* 1993, 90:10676-10680.
24. Charlesworth D, Wright S: Breeding systems and genome evolution. *Current Opinion in Genetics & Development* 2001, 11(6):685-690.
25. Hungerer KB, Kadereit JW: The phylogeny and biogeography of *Gentiana* L. sect.*Ciminalis* (Adans.) Dumort: a historical interpretation of distribution ranges in the European high mountains. *Perspectives in Plant Ecology, Evolution & Systematics* 1998, 1(1):121-135.
26. Posada D, Crandall KA: MODELTEST: testing the model of DNA substitution. *Bioinformatics* 1998, 14:817-818.
27. Suh Y, Thien LB, Reeve HE, Zimmer EA: Molecular evolution and phylogenetic implications of internal transcribed spacer sequences of ribosomal DNA in Winteraceae. *American Journal of Botany* 1993, 80(9):1042-1055.
28. Sang T, Crawford DJ, Kim S-C, Stuessy TF: Radiation of the endemic genus *Dendroseris*(Asteraceae) on the Juan Fernandez

Islands: Evidence from sequences of the ITS regions of nuclear ribosomal DNA. *American Journal of Botany* 1994, 81(11):1494-1501.

29. Magallon SA: Dating lineages: molecular and paleontological approaches to the temporal framework of clades. *International Journal of Plant Sciences* 2004, 165(4 Suppl):S7-S21.
30. Magallon S, Sanderson MJ: Absolute diversification rates in angiosperm clades. *Evolution* 2001, 55(9):1762-1780.
31. Sanderson MJ, Doyle JA: Sources of error and confidence intervals in estimating the age of angiosperms from rbcL and 18S rDNA data. *Am J Bot* 2001, 88(8):1499-1516.
32. Nichols R: Gene trees and species trees are not the same. *Trends in Ecology & Evolution* 2001, 16(7):358-364.
33. Huelsenbeck JP, Rannala B: Phylogenetic methods come of age: testing hypotheses in an evolutionary context. *Science* 1997, 276(5310):227-232.
34. Felsenstein J: Phylogenies from molecular sequences – inference and reliability. *Annual Review of Genetics* 1988, 22:521-565.
35. Sanderson MJ: r8s: inferring absolute rates of molecular evolution and divergence times in the absence of a molecular clock. *Bioinformatics* 2003, 19(2):301-302.
36. Grant V: Historical development of ornithophily in the western North-American flora. *Proceedings of the National Academy of Sciences of the United States of America* 1994,91(22):10407-10411.
37. Hodges SA, Fulton M, Yang JY, Whittall JB: Verne Grant and evolutionary studies of*Aquilegia*. *New Phytologist* 2004, 161(1):113-120.
38. Hickman JC, ed: *The Jepson manual: higher plants of California*. Berkeley: University of California Press; 1993.
39. Mabberley DJ: *The plant-book: a portable dictionary of the vascular plants*. 2nd edition. Cambridge: Cambridge University Press; 1997.
40. Webb C, Donoghue M: Phylomatic. [http://www.phylodiversity.net/phylomatic/] *webcite Tree version R20031202 edn* 2003.
41. APG II: An update of the Angiosperm Phylogeny Group

classification for the orders and families of flowering plants: APG II. *Botanical Journal of the Linnean Society* 2003, 141(4):399-436.

42. Doyle JJ, Doyle JL, Ballenger JA, Dickson EE, Kajita T, Ohashi H: A phylogeny of the chloroplast gene rbcL in the Fabaceae: taxonomic correlations and insights into the evolution of nodulation. *American Journal of Botany* 1997, 84:541-554.
43. Bayer RJ, Starr JR: Tribal phylogeny of the Asteraceae based on two noncoding chloroplast sequences, the trnL intron and trnL/trnF intergenic spacer. *Annals of the Missouri Botanical Garden* 1998, 85(2):242-256.
44. Reeve J, Abouheif E: Phylogenetic Independence.[http://www.biology.mcgill.ca/faculty/abouheif/] *webcite* edition. Version 2.0; 2003.
45. Abouheif E: A method for testing the assumption of phylogenetic independence in comparative data. *Evolutionary Ecology Research* 1999, 1:895-909.
46. Purvis A, Rambaut A: Comparative analysis by independent contrasts (CAIC): an Apple Macintosh application for analysing comparative data. *Computational and Applied Bioscience* 1995, 11(3):247-251.
47. Swofford DL: *PAUP*: phylogenetic analysis using parsimony ; 4.0b10.* edition. Sunderland, MA: Sinauer Associates; 1998.
48. Wen J, Shi S: A phylogenetic and biogeographic study of *Hamamelis*(Hamamelidaceae), an eastern Asian and eastern North American disjunct genus. *Biochemical Systematics & Ecology* 1999, 27(1):55-66.
49. Manos PS: Systematics of *Nothofagus* (Nothofagaceae) based on rDNA spacer sequences (ITS): Taxonomic congruence with morphology and plastid sequences. *American Journal of Botany* 1997, 84(8):1137-1155.
50. Leskinen E, Alstrom-Rapaport C: Molecular phylogeny of Salicaceae and closely related Flacourtiaceae: evidence from 5.8 S, ITS 1 and ITS 2 of the rDNA. *Plant Systematics & Evolution* 1999, 215(1–4):209-227.
51. Wen J: Internal transcribed spacer phylogeny of the Asian and

eastern North American disjunct *Aralia* sect.*Dimorphanthus* (Araliaceae) and its biogeographic implications. *International Journal of Plant Sciences* 2000, 161(6):959-966.

52. Boehle U-R, Hilger HH, Martin WF: Island colonization and evolution of the insular woody habit in *Echium* L.(Boraginaceae). *Proceedings of the National Academy of Sciences of the United States of America* 1996,93(21):11740-11745.
53. Savard L, Michaud M, Bousquet J: Genetic diversity and phylogenetic relationships between birches and alders using ITS, 18S rRNA, and rbcL gene sequences. *Molecular Phylogenetics & Evolution* 1993, 2(2):112-118.
54. Li J, Alexander J, Ward T, Del Tredici P, Nicholson R: Phylogenetic relationships of Empetraceae inferred from sequences of chloroplast gene matK and nuclear ribosomal DNA ITS region. *Molecular Phylogenetics & Evolution* 2002, 25(2):306-315.
55. Vargas P, Morton CM, Jury SL: Biogeographic patterns in Mediterranean and Macaronesian species of *Saxifraga* (Saxifragaceae) inferred from phylogenetic analyses of ITS sequences. *American Journal of Botany* 1999, 86(5):724-734.
56. Xiang Q-Y, Crawford DJ, Wolfe AD, Tang Y-C, Depamphilis CW: Origin and biogeography of *Aesculus* L. (Hippocastanaceae): A molecular phylogenetic perspective. *Evolution* 1998, 52(4):988-997.
57. Thulin M, Lavin M: Phylogeny and biogeography of the *Ormocarpum* group (Fabaceae): A new genus *Zygocarpum* from the Horn of Africa region. *Systematic Botany* 2001, 26(2):299-317.
58. Richardson JE, Weitz FM, Fay MF, Cronk QCB, Linder HP, Reeves G, Chase MW:Phylogenetic analysis of *Phylica* L. (Rhamnaceae) with an emphasis on island species: evidence from plastid trnL-F and nuclear internal transcribed spacer (ribosomal) DNA sequences. *Taxon* 2001, 50(2):405-427.
59. Baum DA, Small RL, Wendel JF: Biogeography and floral evolution of Baobabs (*Adansonia*, Bombacaceae) as inferred from multiple data sets. *Systematic Biology* 1998, 47(2):181-207.
60. Schmidt GJ, Schilling EE: Phylogeny and biogeography of *Eupatorium* (Asteraceae: Eupatorieae) based on nuclear ITS

sequence data. *American Journal of Botany* 2000, 87(5):716-726.

61. Baldwin BG, Sanderson MJ: Age and rate of diversification of the Hawaiian silversword alliance (Compositae). *Proceedings of the National Academy of Sciences of the United States of America* 1998,95(16):9402-9406.
62. Lavin M, Wojciechowski MF, Gasson P, Hughes C, Wheeler E: Phylogeny of robinioid legumes (Fabaceae) revisited: *Coursetia* and *Gliricidia* recircumscribed, and a biogeographical appraisal of the Caribbean endemics. *Systematic Botany* 2003, 28(2):387-409.
63. Kass E, Wink M: Molecular phylogeny and phylogeography of *Lupinus* (Leguminosae) inferred from nucleotide sequences of the rbcL gene and ITS 1+2 regions of rDNA. *Plant Systematics and Evolution* 1997, 208(3–4):139-167.
64. Wojciechowski MF, Sanderson MJ, Hu J-M: Evidence on the monophyly of *Astragalus*(Fabaceae) and its major subgroups based on nuclear ribosomal DNA ITS and chloroplast DNA trnL intron data. *Systematic Botany* 1999, 24(3):409-437.
65. Jobst J, King K, Hemleben V: Molecular evolution of the internal transcribed spacers (ITS1 and ITS2) and phylogenetic relationships among species of the family Cucurbitaceae. *Molecular Phylogenetics & Evolution* 1998, 9(2):204-219.
66. Verboom GA, Linder HP, Stock WD: Phylogenetics of the grass genus *Ehrharta* : evidence for radiation in the summer-arid zone of the South African Cape. *Evolution* 2003, 57(5):1008-1021.
67. von Hagen KB, Kadereit JW: The phylogeny of *Gentianella* (Gentianaceae) and its colonization of the southern hemisphere as revealed by nuclear and chloroplast DNA sequence variation. *Organisms Diversity & Evolution* 2001, 1(1):61-79.
68. Wendel JF, Schnabel A, Seelanan T: An unusual ribosomal DNA sequence from*Gossypium gossypioides* reveals ancient, cryptic, intergenomic introgression. *Molecular Phylogenetics & Evolution* 1995, 4(3):298-313.
69. Sang T, Crawford DJ, Stuessy TF, Silva-O M: ITS sequences and phylogeny of the genus*Robinsonia* (Asteraceae). *Systematic Botany* 1995, 20(1):55-64.
70. Zhang L-B, Comes HP, Kadereit JW: Phylogeny and quaternary

history of the European montane/alpine endemic *Soldanella* (Primulaceae) based on ITS and AFLP variation. *American Journal of Botany* 2001, 88(12):2331-2345.

71. Rahn K: A phylogenetic study of the Plantaginaceae. *Bot J Linn Soc* 1996, 120:145-98.

Chapter 8

Molecular Evolution and the Latitudinal Biodiversity Gradient

E J Dowle[1], M Morgan-Richards[1], and S A Trewick[1]

[1]Ecology Group, Massey University, Palmerston North, New Zealand

ABSTRACT

Species density is higher in the tropics (low latitude) than in temperate regions (high latitude) resulting in a latitudinal biodiversity gradient (LBG). The LBG must be generated by differential rates of speciation and/or extinction and/or immigration among regions, but the role of each of these processes is still unclear. Recent studies examining differences in rates of molecular evolution have inferred a direct link between rate of molecular evolution and rate of speciation, and postulated these as important drivers of the LBG. Here we review the molecular genetic evidence and examine the factors that might be responsible for differences in rates of molecular evolution. Critical to this is the directionality of the relationship between speciation rates and rates of molecular evolution.

INTRODUCTION

'Animal life is, on the whole, far more abundant and more varied within the tropics than in any other part of the globe, and a great number of peculiar groups are found there which never extend into temperate regions'—AR Wallace, 1876.

One of the most striking biogeographic patterns on the planet is the uneven latitudinal distribution of biodiversity (Figure 1). A similar trend of high and low diversity is documented in all major groups of organisms, on the land and in the sea (Hillebrand, 2004).

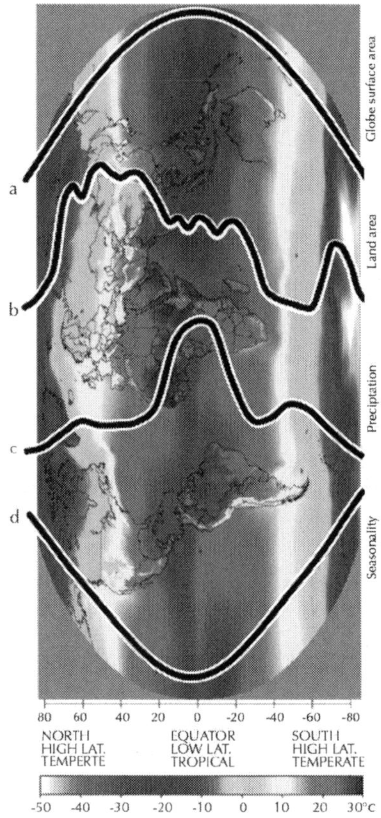

Figure 1: A latitudinal biodiversity gradient (LBG) has been identified in almost all organisms that have been investigated, on land and in the sea. The gradient involves high species' numbers near the equator (at low latitudes)

and lower numbers of species at high latitudes. The question of why and how this pattern has arisen has persisted for more than 150 years. Several environmental parameters form latitudinal gradients reflecting the shape of the planet, its rotation and orientation to the sun. Although the globes' surface area (geometric calculation of a dome) is evenly graded with latitude (a), the extent of land area (b), ocean and variation in bathymetry are not graded evenly. Associated wind and ocean currents further complicate the distribution of environmental variables. Average marine and terrestrial temperatures reduce with increasing latitude (background), but precipitation (c) (L.D.Roper Virginia Poly. Inst. and State University) does not have a linear relationship and seasonality (for example, in terms of annual variance of insolation data fromhttp://www.applet-magic.com/insolation.htm) increases (d). Correlation of these traits with the LBG does not demonstrate causation. Temperature map produced by R.A Rhode,http://en.wikipedia.org/wiki/File: Annual_Average_Temperature_Map.jpg. Details are available from the authors.

The evolutionary and ecological factors responsible for the pattern usually referred to as the latitudinal biodiversity gradient (LBG) have been hotly debated over the past 30 years (Rohde, 1992; Gaston, 2000; Hawkins, 2004; Losos, 2008;Erwin, 2009; Gillman *et al.*, 2009). Interest in the drivers behind the LBG has expanded from biogeographers (starting with Wallace) to include ecologists, evolutionary biologists (in particular evolutionary geneticists), physiologists and palaeontologists (Rosenzweig, 1995; Allen *et al.*, 2002; Davies *et al.*, 2004;Jablonski *et al.*, 2006), however, no consensus has been reached. There is general agreement that variation in the total number of species among regions must result from variation in one or more of three basic phenomena underpinning the diversity of life on the earth: speciation rates, extinction rates and immigration rates. Regional biotas are the product of the interaction of these three processes, but understanding what factors drive these to form the LBG is problematic (Jablonski *et al.*, 2006). How far have we progressed in developing an explanation since Alfred Wallace's observation more than 130 years ago? Here we focus on recent molecular genetic studies to see whether this field, which has contributed so much to evolutionary research, has helped our understanding of the LBG.

Speciation

Differences in speciation rates have been the major focus of geneticists, ecologists and palaeontologists seeking to explain the LBG. The tropics are referred to as a cradle of diversity with high speciation rates inferred from observed high species' diversity (Stebbins, 1974; Chown and Gaston, 2000). Palaeontological data allow comparison of direct counts of species' origins (speciation) in past tropical and temperate areas and have provided the most compelling evidence for differing speciation rates (Jablonski *et al.,* 2006). Molecular data are increasingly applied to the study of speciation rates, but these analyses are not always directed specifically at the LBG (Lancaster, 2010; Lanfear*et al.,* 2010a). Although there is compelling evidence of relatively high rates of speciation in the tropics, the underlying driver(s) of this have not been identified (Martin and Mckay, 2004; Allen and Gillooly, 2006; Jablonski *et al.*, 2006;Mittelbach *et al.*, 2007; Krug *et al.*, 2009; Condamine *et al.*, 2012).

Extinction

A higher rate of extinction at higher latitudes (away from the equator) was the explanation offered by Wallace for the formation of the LBG (Wallace, 1876). Hypotheses focusing on extinction rates have often centred on the putative effects of climatic extremes in the past (for example, during the Pleistocene), which may have been felt most intensely at higher latitudes (Figure 1). In the tropics, a more uniform climate over time may have allowed the accumulation of species and these regions have thus been referred to as museums (Wallace, 1876; Wright, 1983; Guo and Ricklefs, 2000). Unfortunately, the evidence for extinction and its effect is striking by its absence in discussions on this topic, reflecting a general problem with the quantification and accommodation of extinction in biogeography (Crisp *et al.,* 2011). Only palaeontology has provided direct counts of extinction and some comparison of extinction rates, mostly in marine invertebrates. It is well recognised that climate cycling caused extinction in temperate areas during the Pleistocene, but over a longer time frame the evidence is more equivocal. Few studies find strong support for differences in extinction rates between latitudes, and the general inference is that extinction alone cannot explain the LBG (Hawkins *et al.*, 2006; Jablonski *et al.*, 2006; Martin*et al.*, 2007; Valentine *et al.*, 2008).

Immigration

Although immigration is central to some biogeographic models (Macarthur and Wilson, 1967), it has been considered, by some, to be too weak a force to be important in as large scale a trend as the LBG (Gaston, 2000). To be relevant in this context, immigration needs to involve the permanent range change of a tax on resulting in its absence from its original range. Although evidence of large-scale species-range movement has been reported in marine bivalves, terrestrial species may be more range limited (Jablonski *et al.*, 2006; Martin *et al.*, 2007; Mittelbach*et al.*, 2007). Niche conservation has been interpreted by some as indicating a limited capacity in many plant and animal groups for substantial range (habitat) shifts (for example, Wiens and Graham 2005). Immigration has yet to be linked to molecular evolution directly, although range changes could have important implications for the assumptions within many molecular approaches. In particular, the assumption that the current range midpoint of a species represents the midpoint through most of its evolutionary history is questionable.

Topof page

THE PROCESS BEHIND THE PATTERN

'The causes of these essentially tropical features are not to be found in the comparatively simple influences of solar light and heat, but rather in the uniformity and permanence with which these and all other terrestrial conditions have acted; neither varying prejudicially throughout the year, nor having undergone any important change for countless past ages'.—AR Wallace, 1876

The LBG is an observable pattern that, as is typical of biogeographic patterns, does not in itself provide information about the processes involved in its formation (Figures 1 and 2). Latitude values provide a convenient scale for graphing biodiversity patterns but do not directly express the driving force(s) of the gradient (Zapata *et al.*, 2003; Currie and Kerr, 2008). In recent work, temperature has become a focal putative driver replacing an earlier emphasis on habitat area.

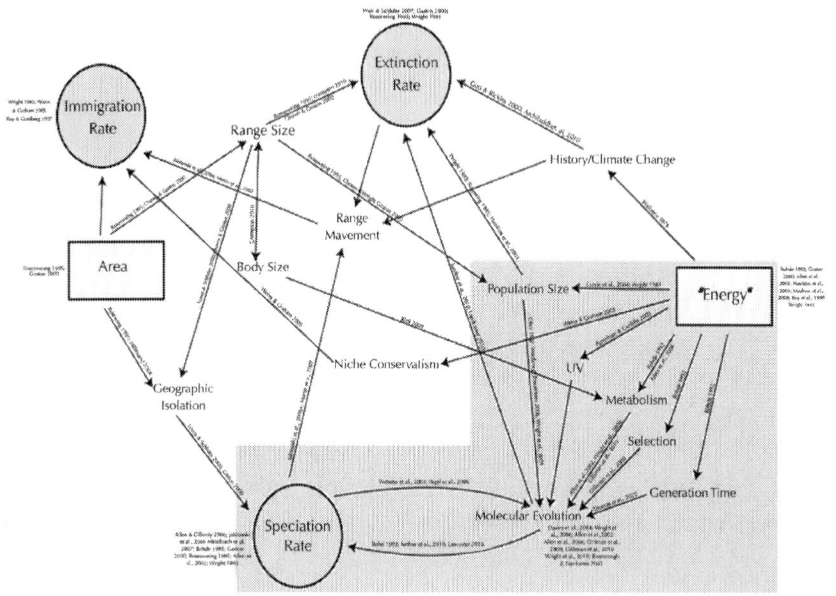

Figure 2: A multitude of hypotheses seek to explain the latitudinal biodiversity gradient (LBG). Each hypothesis aim to relate the LBG to processes that influence one or more of the three components controlling regional species diversity: immigration, extinction and speciation. The two key environmental parameters most frequently cited as underlying drivers of the LBG are area and energy (generally measured in the form of temperature). The inferred importance and directionality of effects varies among taxa, researchers and methods applied, with no clear consensus. Here we illustrate a subset of theories on the formation of the gradient that have gathered some empirical support (citations in figure) and, in particular, those that relate to molecular evolution (shaded area). A link between speciation rate and the rate of molecular evolution is widely shown, but which is the driver of the other is equivocal. A latitudinal gradient in available energy may influence many traits but these are only tentatively associated with rates of extinction and/or speciation.

Palaeontological studies find evidence that environmental temperature is correlated with biodiversity through time in marine molluscs and protists (Crame, 2002; Mayhew *et al.*, 2008; but see

Crampton *et al.*, 2006; where no link was found for molluscs). Most studies find the strongest correlation with temperature (or in some cases ultraviolet radiation, UV) when comparing species richness with a number of possible variables (for example, area, mid-domain effects; Clarke and Gaston, 2006). Several mechanisms have been proposed for how higher temperature might lead to greater species diversity by influencing the speed of speciation, species' physiological tolerances and extinction rates. The 'evolutionary speed' hypothesis has thus become the explanation for the biodiversity gradient preferred by many molecular biologists.

The Evolutionary Speed Hypothesis

Rohde (1992) described the evolutionary speed hypothesis to explain the LBG in terms of a difference in speciation rates; speciation rates being influenced by a latitudinal disparity in rates of molecular evolution and selection. Any differences in rates of molecular evolution between latitudes are considered to be the products of the different environmental conditions that individuals experience at different latitudes. Differing rates of molecular evolution might influence speciation rates if the latter are determined, at least partially, by the random accumulation of genetic differences (mutation-order speciation) and natural selection (ecological speciation; Schluter, 2009; Lancaster, 2010). Allopatric speciation via genetic drift (mutation-order speciation) might result in the formation of new species just as rapidly as ecological speciation, however, the relative influence of these two processes in speciation is not known (Schluter, 2009; Nosil and Flaxman, 2011). Ecological speciation is the more readily studied of the two, and can accommodate difficulties of gene flow (Rundle *et al.*, 2000; Funk *et al.*, 2006;Langerhans *et al.*, 2007; Schluter, 2009). Mutation-order speciation might contribute to further separation of populations already partitioned by natural selection. Alternatively, it could result in faster reproductive isolation by genetic drift in geographically isolated (allopatric) populations perhaps via a Dobzhansky-Muller process (Schluter, 2009; Nosil and Flaxman, 2011).

A faster mutation rate could increase the rate of speciation (for example, seeLancaster, 2010), if mutation-order speciation contributes significantly to overall speciation rate. This presumes that all other evolutionary processes (for example, competition, character displacement, reproductive cohesion, adaptation) are limited only by mutation rate. Central to the biodiversity debate, is the conundrum: does speciation result in accelerated molecular evolution (Webster *et al.*, 2003; Pagel *et al.*, 2006) or does accelerated molecular evolution increase the chance of speciation? That even this fundamental point is subject to debate highlights just how little we understand the role of genetic variation in the LBG and speciation generally.

The evolutionary speed hypothesis predicts an elevated rate of molecular evolution towards the tropics and that this variation directly effects relative rates of speciation. A latitudinal difference in rates of molecular evolution (faster in the tropics) has indeed been found in a range of organisms using sister species comparisons of plants, fish, frogs, foraminifera and mammals (Davies *et al.*, 2004;Allen *et al.*, 2006; Wright *et al.*, 2006; Gillman *et al.*, 2009, 2010; Wright *et al.*,2010, 2011), although the relationship for mammals may be weaker than originally reported (Weir and Schluter, 2011; Figure 3). Rohde (1992) considered three main explanations for increased rates of molecular evolution: shorter generation times; direct effects of temperature on mutation rate; and acceleration of mutation rates due to acceleration of physiological process along with an acceleration of positive selection owing to the first two. These explanations have since been elaborated and extended to include the postulated influence of: metabolic rates, generation time, UV radiation and population size (Figure 2). Here we examine these four putative drivers of differential rates of molecular evolution. To fully understand how rates of molecular evolution could be influencing the biodiversity gradient, it is important to understand how and why these drivers could affect rates of molecular evolution in the first place, providing context to studies of molecular evolution rate variation among latitudes.

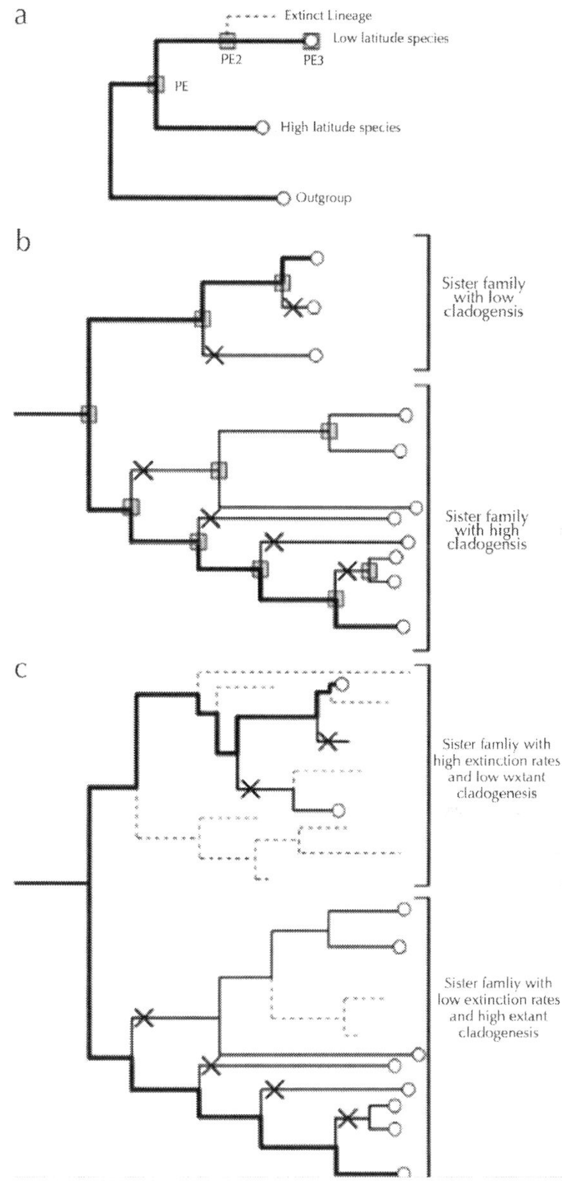

Figure 3: Evidence of variation in the rate of molecular evolution has come from two main sampling strategies that use DNA sequences from extant species (usually represented by a single individual) to infer rate differences among lineages Possible instances of punctuational evolution at nodes (grey boxes), extinct lineages (dashed lines), pruned or un-sampled lineages (X) are shown.

(a) The sister species comparison. Sister species are chosen to represent contrasting distributions; one at low latitude and one at high latitude (temperate versus tropical as in Gillman *et al.*,2009). The placement of the outgroup in the phylogenetic reconstruction provides the crucial evidence for rate variation between the two ingroup lineages. By sampling only one individual per lineage, this method avoids the node-density effect that might influence estimates of molecular evolution rate (Hugall and Lee, 2007). Punctuated equilibrium (PE) predicts that acceleration in rates of molecular evolution is associated with speciation and therefore occurs at nodes. If so, observed differences in rates (different branch lengths) cannot result from PE unless speciation (PE_2) and extinction have occurred only on the branch leading to the tropical species. Hidden (or incipient) speciation (PE_3) could also explain the observed faster rate of molecular evolution if it was occurring more frequently in the tropical lineages than the temperate lineages. (b) Pruning large data sets. Species are selected from sister lineages with different species numbers to examine whether or not lineages with higher cladogenesis have faster rates (for example, Lanfear *et al.*, 2010a). Relatively large data sets are pruned (X) to a pair of contrasting lineages, if more sequence data are available for one family than another, to avoid the node density effect. (c) Pruning large data sets with extinction. Lineage extinction (dashed lines) undermines confidence in inferences about the directionality of effects. Surviving species only provide estimates of net-diversification (speciation−extinction), not rates of speciation *per se*. Insufficient knowledge of extant lineages produces the same effect as extinction suggesting that this method would only work for well-studied groups.

Metabolic Rate Variation Driving Differential Rates of Molecular Evolution

Work on metabolism and its influence on the LBG started as part of the 'metabolic theory of ecology' (Allen *et al.*, 2002; Gillooly and Allen, 2007). It has been suggested that the LBG arose in response to a latitudinal gradient in kinetic energy influencing rates of molecular evolution in ectotherms (Allen and Gillooly, 2006; Gillooly and Allen, 2007). Under this model, ectotherms in warmer areas have an increased metabolic rate governing consumption of oxygen and production of oxygen-free radicals that potentially increase mutation rate by damaging DNA (Allen *et al.*, 2006; Gillooly *et al.*, 2007). Comparing species with different metabolic rates is a proxy for comparison of different rates of oxygen-free radical formation (and inferred mutation), which until recently could not easily be measured directly. New sequencing technology now allows for comparison of whole genomes of parents and their offspring, which can be used to estimate *de novo* mutation rates of species (see below). In comparison to other putative divers, the concept of metabolic rates and the LBG yields clearly testable predictions, but has been heavily contested since its origin (Algar *et al.*, 2007;Hawkins *et al.*, 2007a, 2007b; De Castro and Gaedke, 2008; Irlich *et al.*, 2009).

Unexpectedly, elevated rates of molecular evolution in the tropics compared with temperate regions have also been observed in some endotherms (for example, mammals; Gillman *et al.*, 2009), which was not predicted by the original metabolic hypothesis (Storch, 2003; Gillman *et al.*, 2009). Metabolic rates in endotherms are closely linked to body size, with larger bodies having higher absolute metabolic rates but lower rates per unit mass. As body size in endotherms tends to be greater at higher latitudes (Bergmann's rule), a counter gradient decrease in basal metabolic rate per unit mass with increased latitude is suggested (Ashton *et al.*,2000). But Gillman *et al.* (2009) found that even when body size of mammals was accounted for, the rate of molecular evolution still increased towards the tropics. There is, however, evidence from comparative studies that basal metabolic rate is de-coupled from molecular evolution in the bird and mammal mitochondrial DNA (mtDNA) that are widely used in these studies (Lanfear *et al.*, 2007; Nabholz *et al.*, 2009; Galtier *et al.*, 2009a). Thus, the proxy for oxygen-free radicals (metabolic rate) may inform little on rates of mutation or molecular evolution, and we are no closer to understanding the driver of variation in the rate of molecular evolution and the formation of the LBG.

Generation Time and Longevity Driving Differential Rates of Molecular Evolution

Generation times are generally assumed to be shorter in tropical organisms than their temperate counterparts. Having a short generation time could cause lineages to accumulate more mutations in a given time period compared with lineages with long generation times (Thomas *et al.*, 2010). Although under-examined, the predicted correlation between generation time and rates of molecular evolution has been observed in some groups of invertebrates, flowering plants, host–parasite systems and mammalian nuclear sequences (Nikolaev *et al.*, 2007; Smith and Donoghue, 2008; Welch *et al.*, 2008; Thomas *et al.*, 2010), but not for other plants or mammal mtDNA (Whittle and Johnston, 2003; Nabholz *et al.*, 2008). Ohta (1992) suggested that variation in generation time and population size was the cause of the discrepancy between observed rates of protein and nucleotide evolution.

Plants (especially large, long-lived species) can be problematic subjects with regard to generation time, because mutations in reproductive tissue are also influenced by mutagenesis of somatic cells (Gaut *et al.*, 2011). Little direct testing of generation times across latitudes has been undertaken, probably because the basic biology of many tropical species is not known. Linked to this problem is the issue of longevity, which refers to the number of cell cycles an individual undergoes rather than the life expectancy of the organism. This is another life history hypothesis that has yet to be implicated in the LBG, but has a similar relationship to rates of molecular evolution as generation time. Longer-lived species (with more cell cycles per individual) are expected to be more disadvantaged by higher mutation rate, as premature aging could result from a high somatic DNA mutation rate (Nabholz *et al.*, 2008, 2009).

UV Radiation Driving Differential Rates of Molecular Evolution

UV is a mutagen and as a component of insolation is most constant and intense at the equator (Figure 1; Rastogi *et al.*, 2010). UV intensity has been linked to the LBG by the prediction that an increased mutation rate with higher exposure to UV (Bromham and Cardillo, 2003; Davies *et al.*, 2004; Flenley, 2011) may be correlated with speciation rate. The proposed role of UV as a producer of heritable mutations is perhaps more straightforward in animals where exposure to UV could occur during embryonic development (Epel *et al.*, 1999; Maurer *et al.*, 2011) as this could result in germline mutations. In plants, UV has been proposed as a potential mutagen of pollen grains and experimental studies have shown its effects (Ahmad *et al.*, 1991; Flenley, 2011). In mammals, a role for UV in germline mutations is less obvious but not incomprehensible (Wilson Sayres and Makova, 2011). UV levels are generally higher in the tropics than in temperate regions, however, UV radiation, unlike temperature, does not decrease with altitude and may even increase (Korner, 2007; Barry, 2008). A number of studies exploring differences in rates of molecular evolution have considered sister species pairs from different altitudes (Wright *et al.*, 2006; Gillman *et al.*, 2009, 2010; Wright *et al.*, 2010). In all cases (plants, mammals and amphibians), inferred rates of molecular evolution were higher at lower altitudes, contrary to the prediction of UV as a biodiversity driver.

Population Size Driving Differential Rates of Molecular Evolution

The effects of population size are complex and important as they may interact with other putative drivers of LBG (Charlesworth, 2009). There is an extensive literature relating to the influence of population size, comprising many contrasting views and interpretations. Population size is often postulated as a component of LBG hypotheses, but is variously assumed to be larger and smaller in the tropics, although evidence for either view is limited (Wright, 1983; Currie *et al.*, 2004;Gillman *et al.*, 2009). Land area of islands is commonly used as a proxy for population size in studies of rates of molecular evolution in terrestrial organisms (Woolfit and Bromham, 2005; Wright *et al.*, 2009), an assumption founded in island biogeography theory (Macarthur and Wilson, 1967).

Although there is some limited evidence for latitudinal gradients in population size on a global scale, the topic is controversial because of the contested influence of population size on rates of molecular evolution. This is frequently confounded by a lack of clarity about what estimate of molecular evolution is being used: relative branch length or population variation. Theory predicts that large populations will have more genetic diversity at any one time compared with small populations (Ellstrand and Elam, 1993). This is because smaller populations are more sensitive to genetic drift so that mutations are likely to go to fixation more quickly resulting in lower net genetic variation at the population level. Furthermore, the fewer individuals there are in absolute terms the fewer mutations there will be in the population. Thus, large populations might contain the genetic diversity required for ecological speciation. The connection between population size and relative branch length is less clear (Johnson and Seger, 2001; Woolfit and Bromham, 2005;Nikolaev *et al.*, 2007; Wright *et al.*, 2009). Branch lengths in a phylogeny, expressing local rates of molecular evolution, are influenced by mutation rate and fixation rate. Under neutral theory the ratio of non-synonymous nucleotide mutations (dN) to synonymous nucleotide mutations (dS) will be equal to their respective proportions in the genome (Kimura, 1968). Therefore, the rate of molecular evolution would not be influenced by population size under a neutral model. Synonymous nucleotide mutations do not result in amino-acid substitutions and are usually neutral with respect to natural selection, having no phenotypic effect. In contrast, non-synonymous nucleotide mutations result in amino-acid substitutions and are therefore frequently not neutral. Almost all studies investigating population size and the rate of molecular evolution violate the assumption of neutrality by studying areas of the genome that are known to be constrained by selection (for example, coding regions of mitochondrial genomes such as COI and Cyt*b*). Nearly-neutral theory was developed in response to observations of a mismatch between inferred short-term and long-term mutation rates, and predicts that slightly deleterious mutations, which would be eliminated in large populations, may tend to persist longer in smaller populations where purifying selection is more relaxed (Ohta, 1992). Thus, the nearly-neutral model predicts more non-synonymous substitutions in small populations, but synonymous substitutions should be independent of population size (Ohta, 1992). This leads to the prediction that dN/dS (ω) will differ between large and small populations, and that changes in population size (expansion and contractions) will, in theory, alter rates of molecular evolution. Some empirical evidence agrees with this model (Johnson and Seger, 2001; Woolfit and Bromham, 2005; Bakewell *et al.*, 2007;Charlesworth and Eyre-Walker, 2007; Kosiol *et al.*, 2008; Petit and Barbadilla, 2009). However, some of these studies have found the results to be weak, with one finding the reverse (dN and dS were both lower in smaller than larger populations; Wright *et al.*, 2009). Explanations proposed for these results include a role for positive selection in the larger populations, positive back-mutations during population expansion and/or linked substitutions (Charlesworth and Eyre-Walker, 2007; Wright *et al.*, 2009; Stoletzki and Eyre-Walker, 2011). Positive selection complicates matters, as the efficiency of selection in larger populations results in the retention of beneficial mutations, whereas in smaller populations beneficial mutations act more like neutral mutations (Charlesworth, 2009). In addition, these studies have tended to use a limited number of markers; often a single-mtDNA sequence, which is subject to the Hill-Robertson effect due to selection and lack of recombination (Charlesworth, 2009). The demographic influence of population size on the gradient of molecular evolution remains unresolved, due to the continued debate and conflicting empirical evidence on the influence of population size on molecular evolution rates and uncertainty about actual population size differences between high and low latitude species.

TOWARDS A SYNTHESIS

Rates of Molecular Evolution and Speciation

Evidence for differential rates of molecular evolution being correlated with latitude are compelling yet they lack one key feature as a model to explain the biodiversity gradient; no study has yet convincingly linked differences in rates of molecular evolution to differences in speciation, extinction or immigration rates (Wright *et al.*, 2006; Gillman *et al.*, 2009, 2010; Wright *et al.*, 2010). Although the evidence for elevated rates of molecular evolution at the tropics is substantial, it remains only a pattern. Many traits that have been linked to the biodiversity gradient, such as niche conservation, similarly describe a pattern (an outcome) rather than the process (driver) involved in generating the gradient (Wiens and Graham, 2005;Losos, 2008; Crisp and Cook, 2012). Rates of molecular evolution and numbers of species show latitudinal clines, but are they directly related to one another, or is a third independent variable implicated? For example, differences in branch length between high- and low-latitude phylogenetically independent angiosperms show higher rates of molecular evolution at low latitude, but this does not correlate with higher speciation rates. Instead, the two rates were each independently linked to an environmental variable (temperature or UV; Davies *et al.*, 2004). Therefore, an indirect link might explain the observed relationship between rates of molecular evolution and biodiversity, or one of these two traits might drive the other. Contrary to the idea that elevated rates of molecular evolution lead to increased speciation, some studies infer that elevated molecular evolution is a result of more speciation, referred to as punctuational evolution

(Pagel *et al.*,2006; Venditti and Pagel, 2010). Thus, the increased rate of molecular evolution in the tropics may be a product of speciation rather than the cause of it (Websteret *al.*, 2003; Pagel *et al.*, 2006).

Evidence of a link between rates of molecular evolution and net-diversification has been reported (Barraclough and Savolainen, 2001; Jobson and Albert, 2002; Eo and Dewoody, 2010; Lancaster, 2010; Lanfear *et al.*, 2010a but see Goldie *et al.*,2011 where no such link was found in mammals). However, none of these studies have conclusively shown that increased rates of molecular evolution are responsible for elevated rates of speciation. Evidence that the rate of molecular evolution is driving speciation rate has been found in plants (Lancaster, 2010), but the inference relies on the approach used to model extinction (Quental and Marshall, 2010). The assumption that the plant clades studied evolved via a 'birth-death' diversification process (Lancaster, 2010) is untested, and it is possible that models of density-dependent speciation (Moran process) or pure birth (Yule process) fit the data better (Nee, 2004). Also, it is unclear whether results from a multicopy nuclear marker that evolves via concerted evolution (Internal Transcribed Spacer (ITS)) can be meaningfully extrapolated to the rest of the genome. An extensive sequence data set from extant birds (Lanfear *et al.*,2010a) identified a link between rates of molecular evolution and net-diversification, but birds are one of the few animal groups where a latitudinal skew in molecular evolution rates has not been conclusively demonstrated (Bromham and Cardillo, 2003), even though birds do contribute to the biodiversity gradient (Hawkins *et al.*, 2003). Lanfear *et al.* (2010a) found that as dS and dN increased in bird lineages so did net-diversification, but they could not find a link between life history traits and an increased rate of molecular evolution. Within their data set there was no association between net-diversification and ω (dN/dS). This suggests that selection and population size were not influencing net-diversification rates, but that the rate of molecular evolution was the most important influence on the total number of species within a clade. Nevertheless, this result does not exclude the possibility that speciation itself accelerated the rate of molecular evolution in these birds, or even that extinction (which contributes to net-diversification) is lower where molecular rate is high (Figure 3).

Molecular Evolution and Extinction

As noted, analyses of molecular genetic data are limited by their inability to directly identify extinction and its effect (Weir and Schluter, 2007; Lancaster, 2010; Quental and Marshall, 2010). Molecular studies are not well suited to examining extinction rates because extinct taxa are not available for sampling. Molecular phylogenies show inferred ancestral relationships among extant (sampled) taxa and do not show how extinction (or failed sampling) influence observed branch lengths. Varying rates of speciation and extinction can leave similar molecular phylogenetic patterns (Rabosky and Lovette, 2008; Crisp and Cook, 2009), so methods such as lineage through time plots may be misleading. It is therefore difficult to disentangle the relative contribution to diversity of speciation and extinction without a good fossil record (Figure 3; Quental and Marshall, 2010).

Evidence from palaeontological studies suggests a combination of lower extinction and higher speciation rates in the tropics has driven the biodiversity gradient (Jablonski *et al.*, 2006; Martin *et al.*, 2007). If rates of molecular evolution are faster in low latitudes as many have found (Davies *et al.*, 2004; Wright *et al.*,2006; Gillman *et al.*, 2009, 2010; Wright *et al.*, 2010), then the relationship identified in mammals and birds by Weir and Schluter (2007) is consistent, but the use of a fixed molecular clock has resulted in misleading inferences of relative speciation and extinction rates. A strict clock rate is inappropriate when it is probable that rates of molecular evolution vary among lineages. Indeed, any variation in diversification rates among lineages, which is likely to be the rule rather than the exception, will interfere with estimations of true extinction rates (Rabosky, 2010). Direct linkage of extinction rates to rates of molecular evolution has also been suggested; increased rates of molecular evolution could reduce extinction risk by increasing the number of beneficial mutations that enable lineages to persist longer (Lanfear *et al.*, 2010a). The effect of mutation rate on extinction is largely unknown, although excessive mutation rates will tend to yield a high genetic load (Butlin *et al.*, 2009; Lancaster, 2010).

Countless Difficulties

'In the one, evolution has had a fair chance; in the other it has had countless difficulties thrown in its way' —AR Wallace, 1876

The future role of molecular research investigating the biodiversity gradient is uncertain as studies have proven to be method-sensitive and open to variable interpretation. Considered individually, many molecular studies appear to provide compelling and relevant evidence, however, the associations they reveal rarely constitute direct evidence of a primary driver. Many studies of the LBG run parallel to one another and cannot be directly compared because they employ very different methods. However, a consensus is developing that there is a correlation between rates of molecular evolution and species diversity; the challenge now is to identify their causal relationship.

Palaeontology has the potential to track diversity trends through time and is therefore well suited to estimate extinction, speciation and immigration rates. A recent test of the influence of seasonality on the biodiversity gradient shows the potential for palaeontology to narrow the field of possible drivers of the LBG (Archibald *et al.*, 2010). Given that sampling of DNA sequence data is limited to extant species and there are known problems with estimating extinction rates from molecular data (Rabosky, 2010), it might be argued that molecular data have a limited role in examining extinction rates, a processes central to the biodiversity gradient debate. However, analytical tools are available if appropriate questions are asked (Nee, 2004), and our ability to incorporate extinction into phylogenetic models is improving (Etienne and Apol, 2009; Morlon *et al.*, 2011; Stadler, 2011). The use of data from measurably evolving populations (for example, in rapidly evolving disease causing microorganisms; and other populations using ancient DNA approaches) will continue to enhance our ability to quantify rates of molecular change and strengthen theoretical developments (Drummond *et al.*, 2003).

Inferences about regional biodiversity and choice of entities for data sampling are both influenced by current taxonomy. In some studies, unrecognised species probably influence the conclusions drawn. It is possible that differences among the extinction and speciation rates inferred in Weir and Schluter (2007) were, in part, artefacts of more intense taxonomic splitting of high-latitude versus low-latitude biota (Tobias *et al.*, 2008). Higher within-species genetic diversity has been found in plants and vertebrates from lower latitudes, and this could be taken to mean that there are more undescribed species in the tropics, or that the tropical taxa have larger population sizes (Martin and Mckay, 2004; Eo *et al.*, 2008). Incorporation of phylogenetic information into estimates of biodiversity (phylogenetic distinctiveness metrics) is increasing (Purvis and Hector, 2000;Davies and Buckley, 2011), but phylogenetic approaches to estimating biodiversity rely in part on the same information (that is, branch length and branching pattern) being used to assess rates of molecular evolution. This circularity weakens inferences about drivers of the LBG that might be further influenced by lineage extinction and local biogeographic history.

Mitochondrial DNA has been used in several studies comparing rates of molecular evolution among lineages with speciation rates. There is some evidence that molecular evolution rates from the mitochondrial genome and the nuclear genome are subject to different pressures that might reflect a difference in proximity to cell metabolic processes or natural selection (Welch *et al.*, 2008). Mitochondrial DNA is frequently used in analyses that assume neutrality in terms of molecular evolution, but this is unlikely to be realistic (for review see Galtier *et al.*, 2009b). Given the metabolic role of mitochondria, the mitochondrial genome could be implicated in climate adaptation, and thus far from neutral in the context of the LBG. This idea has been mooted for human mitochondrial genetics (Ruiz-Pesini *et al.*, 2004; Wallace, 2005; Das, 2006; Pierron *et al.*, 2011), but remains controversial given suggestions that available methods to detect selection within populations are flawed (Kryazhimskiy and Plotkin, 2008). Nuclear 'loci' such as ITS are far from ideal given their complex and poorly understood process of sequence evolution (Elder and Turner, 1995). This could be particularly problematic in plants where hybridisation has an important role in species formation and exchange among lineages (Hegarty and Hiscock, 2005). To date, selection of DNA sequence loci has been strongly influenced by the availability of data and universal PCR primers, a problem which is being alleviated by the use of next generation sequencing (NGS) technology.

FUTURE FOCUS ON THE PROCESS

It is clear that we need approaches that avoid reliance on uncertain proxies. Tests of associations between putative drivers and speciation rates currently rely heavily on substitutes that are assumed to correlate well with the characters of interest. For example, island size is used as a proxy for population size; metabolic rate is used as a proxy for oxygen-free radicals that can influence mutations; average age of first reproduction is a proxy for average species' generation time; and current species distribution is used as a proxy for ancestral species distribution. Many of these correlations first need to be independently verified in the organisms of interest. Also, the design of many studies reflects methodological and sampling limitations, and has often dealt with comparison of just two alternative conditions (for example, big population versus small population, high latitude versus low latitude) rather than analysing gradient or transect data. Although continuous traits are often measured (for example, population size), they are then categorised into two alternatives (for example, big versus small) in order for inferences to be drawn from paired comparisons (for example, sign tests). Mutation rates vary between species, between individuals and within the genome itself, so disentangling all the drivers of mutation rates is unlikely to be easy (Hodgkinson and Eyre-Walker, 2011).

Clarity of assumptions and experimental design and application of methods at the appropriate evolutionary level are essential, and here we suggest some directions that might take us forward.

- Many of the life-history traits that have been implicated as driving rate changes are not independent of one another and work needs to focus on taxa that allow the elimination of some of these variables. The sister species approach has some benefits over the large clade approach, but for life-history traits that cannot be adequately represented across sister species pairs, additional methods will be necessary (see Lanfear *et al.*,2010b for an examination of different approaches to estimate rates from trees). Quantification of life history traits in specific organisms of interest is required to avoid reliance on rules of thumb and generalised concepts that are often based on limited direct evidence. An emphasis on regionally focused examples should enhance opportunities for gathering accurate life-history data, and inclusion of specific geophysical information when accommodating effects of local biogeographic history.
- Until recently, some studies of rates of molecular evolution across the LBG have used single genes from the mtDNA genome and a single branch from each species. NGS technologies now provide multilocus data that improve confidence in estimates of relative rates of molecular evolution across the genome in sister species

comparisons, even for non-model organisms. In the past, when more than one sequence was available per species the shortest branch was routinely chosen (for example, see Wright *et al.,* 2009), even though species usually contain measurable genetic diversity. NGS allows dense population sampling of sister species pairs that allows analysis of the effects of choosing the shortest branch versus other metrics, such as average branch length or random branch selection. Studies, particularly those involving high genetic divergence, may also benefit from an out-group comparison (Hugall and Lee, 2007). Population level studies have the potential to inform on speciation processes at a stage equivalent to the node in a species tree, instead of relying on inferences about past speciation from tree tips.

- For understanding the LBG, determining population size variation between latitudes is important. Thus, a focus on multi-locus data for taxa with sedentary lifestyles would facilitate population density comparisons. If a consistent difference in population size among latitudes exists then the interaction between population size and rates of molecular evolution becomes extremely important to the LBG debate. Multi-locus data sets exist for model organisms (for example, humans and *Drosophila*; Bakewell *et al.,*2007; Kosiol *et al.,* 2008; Petit and Barbadilla, 2009), but it is now possible for substantial tests across the nuclear and mitochondrial genomes functional and non-functional DNA in non-model groups with NGS. However, population size may not be constant through time and methods for inferring past population size that are independent of rates of molecular evolution will be required.
- If speciation occurs predominantly in small isolated populations (Venditti and Pagel, 2010), signal about the size of the populations during speciation may be lost through processes such as back-mutation during subsequent population expansion (Charlesworth and Eyre-Walker, 2007; Lanfear *et al.,*2010a). Simulations using data from model organisms should make it possible to test the predicted population size effect on ω and assess the interaction of ecological speciation and population expansion. It will also be possible to estimate how long after a speciation event the population genetic signature remains detectable. Analysis of recent-past population diversity using ancient DNA methods could enhance modelling of rates of molecular evolution and

population size. Similarly, studies involving ancestral gene reconstruction and resurrection will likely illuminate our understanding of the respective roles of genetic drift and selection on gene evolution (Thornton, 2004).

- Expansion of studies involving lineages represented in the fossil record and neobiota will contribute broadly to meshing information on speciation, extinction and migration with rates of molecular evolution. In particular, such data will help explore the directionality of the relationship between rates of molecular evolution and speciation. Sequencing multiple loci from extant species that are members of fossil-rich lineages will allow extinction rates to be correlated with both speciation and molecular evolution rates. This has the potential to isolate speciation from net-diversification, an important step towards resolving the link between rates of molecular evolution and speciation. Fossil data can be used to constrain molecular clock analyses and allow rate comparisons that are not limited to sister species approaches. Measures of extinction could reveal any relationship between rates of molecular evolution and extinction.
- Sequencing of parents and their offspring allows species-specific estimates of mutation rate, and if the necessary multiple comparisons and multiple tissues are included, the population variance in individual *de novo* mutation rate can be known (Hodgkinson and Eyre-Walker, 2011). Two experimental approaches using this method could contribute directly to the study of LBG. First, the effect of temperature and UV on mutation rate could be assessed using lab-based model organisms in controlled environments. Second, direct comparison of mutation rates of non-model organisms, where rates of molecular evolution have been shown to differ between sister species at separate latitudes, can be obtained from whole-genome parent/offspring data. This would allow discrimination between the two components of molecular evolution, variation in rates of mutation and variation in rates of fixation (Ho *et al.*, 2005). In addition, mining pedigree data sets that exist for applied research in agriculture, horticulture and conservation will allow tests for a link between longevity of individuals within species and mutation rate.

Differences in rates of molecular evolution between global regions are fascinating and warrant much more focused research. But perhaps the priority should be to find the directionality of the link, if any, between speciation and molecular evolution. Without this fundamental information the evolutionary speed hypothesis cannot provide compelling predictions to test putative drivers of the LBG.

ACKNOWLEDGEMENTS

We would like to thank two anonymous reviewers, Robert Lanfear, Roger Butlin and Gillian Gibbs for very helpful comments on earlier versions of this review. The Phoenix group provided useful discussions (evolves@massey.ac.nz), and Massey University and the Royal Society of New Zealand provided funding (Scholarship to EJD; Marsden 04-GNS-021 to MMR).

REFERENCES

1. Ahmad I, Day JP, Macdonald MV, Ingram DS (1991). Haploid culture and UV mutagenesis in rapid-cycling *Brassica napus* for the generation of resistance to chlorsulfuron and *Alternaria brassicicola*. *Ann Botany* 67: 521–519.
2. Algar AC, Kerr JT, Currie DJ (2007). A test of Metabolic Theory as the mechanism underlying broad-scale species-richness gradients. *Global Ecol Biogeogr* 16: 170–178.
3. Allen AP, Brown JH, Gillooly JF (2002). Global biodiversity, biochemical kinetics, and the energetic-equivalence rule. *Science* 297: 1545–1548.
4. Allen AP, Gillooly JF (2006). Assessing latitudinal gradients in speciation rates and biodiversity at the global scale. *Ecol Lett* 9: 947–954.
5. Allen AP, Gillooly JF, Savage VM, Brown JH (2006). Kinetic effects of temperature on rates of genetic divergence and speciation. *Proc Natl Acad SciUSA* 103: 9130–9135.
6. Archibald SB, Bossert WH, Greenwood DR, Farrell BD (2010). Seasonality, the latitudinal gradient of diversity, and Eocene insects. *Paleobiology* 36: 374–398.

7. Ashton KG, Tracy MC, De Queiroz A (2000). Is Bergmann's rule valid for mammals? *Am Nat* 156: 390–415.
8. Bakewell MA, Shi P, Zhang J (2007). More genes underwent positive selection in chimpanzee evolution than in human evolution. *Proc Natl Acad Sci USA* 104: 7489–7494.
9. Barraclough TG, Savolainen V (2001). Evolutionary rates and species diversity in flowering plants. *Evolution* 55: 677–683.
10. Barry RG (2008) *Mountain Weather and Climate*. Cambridge University Press: New York.
11. Bromham L (2011). The genome as a life-history character: why rate of molecular evolution varies between mammal species. *Phil Trans R Soc B Biol Sci* 366: 2503–2513.
12. Bromham L, Cardillo M (2003). Testing the link between the latitudinal gradient in species richness and rates of molecular evolution. *J Evol Biol* 16: 200–207.
13. Butlin R, Bridle J, Schluter D (2009) *Speciation and Patterns of Diversity*. Cambridge University Press: New York.
14. Charlesworth B (2009). Fundamental concepts in genetics: effective population size and patterns of molecular evolution and variation. *Nature reviews. Genetics* 10: 195–205.
15. Charlesworth J, Eyre-Walker A (2007). The other side of the nearly neutral theory, evidence of slightly advantageous back-mutations. *Proc Natl Acad Sci USA* 104: 16992–16997.
16. Chown SL, Gaston KJ (2000). Areas, cradles and museums: the latitudinal gradient in species richness. *Trends Ecol Evol* 15: 311–315.
17. Clarke A, Gaston KJ (2006). Climate, energy and diversity. *Proc R Soc B Biol Sci* 273: 2257–2266.
18. Condamine FL, Sperling FA, Wahlberg N, Rasplus J-Y, Kergoat GJ (2012). What causes latitudinal gradients in species diversity? Evolutionary processes and ecological constraints on swallowtail biodiversity. *Ecol Lett*15: 267–277.
19. Crame JA (2002). Evolution of taxonomic diversity gradients in the marine realm: a comparison of Late Jurassic and Recent bivalve faunas.*Paleobiology* 28: 184–207.
20. Crampton JS, Foote M, Beu AG, Maxwell PA, Cooper RA, Matcham I *et al*(2006). The ark was full! Constant to declining

Cenozoic shallow marine biodiversity on an isolated midlatitude continent. *Paleobiology* 32: 509–532.

21. Crampton JS, Cooper RA, Beu AG, Foote M, Marshall BA (2010). Biotic influences on species duration: interactions between traits in marine molluscs. *Paleobiology* 36: 204–223.
22. Crisp MD, Cook LG (2009). Explosive radiation or cryptic mass extinction? Interpreting signatures in molecular phylogenies. *Evolution* 63: 2257–2265.
23. Crisp MD, Cook LG (2012). Phylogenetic niche conservatism: what are the underlying evolutionary and ecological causes? *New Phytol* 196: 681–694.
24. Crisp MD, Trewick SA, Cook LG (2011). Hypothesis testing in biogeography.*Trends Ecol Evol* 26: 66–72.
25. Currie DJ, Kerr JT (2008). Tests of the mid-domain hypothesis: A review of the evidence. *Ecol Monogr* 78: 3–18.
26. Currie DJ, Mittelbach GG, Cornell HV, Field R, Guegan JF, Hawkins BA *et al*(2004). Predictions and tests of climate-based hypotheses of broad-scale variation in taxonomic richness. *Ecol Lett* 7: 1121–1134.
27. Das J (2006). The role of mitochondrial respiration in physiological and evolutionary adaptation. *BioEssays* 28: 890–901.
28. Davies TJ, Buckley LB (2011). Phylogenetic diversity as a window into the evolutionary and biogeographic histories of present-day richness gradients for mammals. *Phil Trans R Soc B Biol Sci* 366: 2414–2425.
29. Davies TJ, Savolainen V, Chase MW, Moat J, Barraclough TG (2004). Environmental energy and evolutionary rates in flowering plants. *Proc R Soc Lond Series B Biol Sci* 271: 2195–2200.
30. De Castro F, Gaedke U (2008). The metabolism of lake plankton does not support the metabolic theory of ecology. *Oikos* 117: 1218–1226.
31. Drummond AJ, Pybus OG, Rambaut A, Forsberg R, Rodrigo AG (2003). Measurably evolving populations. *Trends Ecol Evol* 18: 481–488.
32. Elder JF, Turner BJ (1995). Concerted evolution of repetitive DNA sequences in eukaryotes. *Q Rev Biol* 70: 297–320.

33. Ellstrand NC, Elam DR (1993). Population genetic consequences of small population size: implications for plant conservation. *Annu Rev Ecol Syst* 24: 217–242.
34. Eo SH, Dewoody JA (2010). Evolutionary rates of mitochondrial genomes correspond to diversification rates and to contemporary species richness in birds and reptiles. *Proc R Soc B Biol Sci* 277: 3587–3592.
35. Eo SH, Wares JP, Carroll JP (2008). Population divergence in plant species reflects latitudinal biodiversity gradients. *Biol Lett* 4: 382–384.
36. Epel D, Hemela K, Shick M, Patton C (1999). Development in the floating world: defenses of eggs and embryos against damage from UV radiation. *Am Zool* 39: 271–278.
37. Erwin DH (2009). Climate as a driver of evolutionary change. *Curr Biol* 19: R575–R583.
38. Etienne RS, Apol MEF (2009). Estimating speciation and extinction rates from diversity data and the fossil record. *Evolution* 63: 244–255.
39. Flenley JR (2011). Why is pollen yellow? And why are there so many species in the tropical rain forest? *J Biogeogr* 38: 809–816.
40. Funk DJ, Nosil P, Etges WJ (2006). Ecological divergence exhibits consistently positive associations with reproductive isolation across disparate taxa. *Proc Natl Acad Sci USA* 103: 3209–3213.
41. Galtier N, Jobson RW, Nabholz B, Glemin S, Blier PU (2009a). Mitochondrial whims: metabolic rate, longevity and the rate of molecular evolution. *Biol Lett* 5: 413–416.
42. Galtier N, Nabholz B, Glémin S, Hurst GDD (2009b). Mitochondrial DNA as a marker of molecular diversity: a reappraisal. *Mol Ecol* 18: 4541–4550.
43. Gaston KJ (2000). Global patterns in biodiversity. *Nature* 405: 220–227.
44. Gaut B, Yang L, Takuno S, Eguiarte L (2011). The patterns and causes of variation in plant nucleotide substitution rates. *Annu Rev Ecol Syst* 42: 245–266.
45. Gillman LN, Keeling DJ, Gardner RC, Wright SD (2010). Faster evolution of highly conserved DNA in tropical plants. *J Evol Biol* 23: 1327–1330.

46. Gillman LN, Keeling DJ, Ross HA, Wright SD (2009). Latitude, elevation and the tempo of molecular evolution in mammals. *Proc R Soc B Biol Sci* 276: 3353–3359.
47. Gillooly JF, Allen AP (2007). Linking global patterns in biodiversity to evolutionary dynamics using metabolic theory. *Ecology* 88: 1890–1894.
48. Gillooly JF, Mccoy MW, Allen AP (2007). Effects of metabolic rate on protein evolution. *Biol Lett* 3: 655–659.
49. Goldie X, Lanfear R, Bromham L (2011). Diversification and rate of molecular evolution: no evidence of a link in mammals. *BMC Evol Biol* 11: 286.
50. Guo QF, Ricklefs RE (2000). Species richness in plant genera disjunct between temperate eastern Asia and North America. *Bot J Linn Soc* 134: 401–423.
51. Hawkins BA (2004). Are we making progress toward understanding the global diversity gradient? *Basic Appl Ecol* 5: 1–3.
52. Hawkins BA, Albuquerque FS, Araujo MB, Beck J, Bini LM, Cabrero-Sanndo FJ*et al* (2007a). A global evaluation of metabolic theory as an explanation for terrestrial species richness gradients. *Ecology* 88: 1877–1888.
53. Hawkins BA, Diniz-Filho JAF, Jaramillo CA, Soeller SA (2006). Post-Eocene climate change, niche conservatism, and the latitudinal diversity gradient of New World birds. *J Biogeogr* 33: 770–780.
54. Hawkins BA, Diniz-Filho JA, Bini LM, Araujo MB, Field R, Hortal J *et al*(2007b). Metabolic theory and diversity gradients: Where do we go from here? *Ecology* 88: 1898–1902.
55. Hawkins BA, Field R, Cornell HV, Currie DJ, Guegan JF, Kaufman DM *et al*(2003). Energy, water, and broad-scale geographic patterns of species richness. *Ecology* 84: 3105–3117.
56. Hegarty MJ, Hiscock SJ (2005). Hybrid speciation in plants: new insights from molecular studies. *New Phyt* 165: 411–423.
57. Hillebrand H (2004). On the generality of the latitudinal diversity gradient.*Am Nat* 163: 192–211.
58. Ho SYW, Phillips MJ, Cooper A, Drummond AJ (2005). Time dependency of molecular rate estimates and systematic overestimation of recent divergence times. *Mol Biol Evol* 22: 1561–1568.

59. Hodgkinson A, Eyre-Walker A (2011). Variation in the mutation rate across mammalian genomes. *Nat Rev Genet* 12: 756–766.
60. Hugall AF, Lee MSY (2007). The likelihood node density effect and consequences for evolutionary studies of molecular rates. *Evolution* 61: 2293–2307.
61. Irlich UM, Terblanche JS, Blackburn TM, Chown SL (2009). Insect rate-temperature relationships: environmental variation and the metabolic theory of ecology. *Am Nat* 174: 819–835.
62. Jablonski D, Roy K, Valentine JW (2006). Out of the tropics: evolutionary dynamics of the latitudinal diversity gradient. *Science* 314: 102–106.
63. Jobson RW, Albert VA (2002). Molecular rates parallel diversification contrasts between carnivorous plant sister lineages. *Cladistics-the Int J Willi Hennig Soc* 18: 127–136.
64. Johnson KP, Seger J (2001). Elevated rates of nonsynonymous substitution in island birds. *Mol Biol Evol* 18: 874–881.
65. Kimura M (1968). Evolutionary rate at the molecular level. *Nature* 217: 624–626.
66. Korner C (2007). The use of 'altitude' in ecological research. *Trends Ecol Evol* 22: 569–574.
67. Kosiol C, Vinar T, Da Fonseca RR, Hubisz MJ, Bustamante CD, Nielsen R *et al*(2008). Patterns of positive selection in six mammalian genomes. *PLoS Genet*4: e1000144.
68. Krug AZ, Jablonski D, Valentine JW, Roy K (2009). Generation of Earth's first-order biodiversity pattern. *Astrobiology* 9: 113–124.
69. Kryazhimskiy S, Plotkin JB (2008). The population genetics of dN/dS. *PLoS Genet* 4: e1000304.
70. Lancaster L (2010). Molecular evolutionary rates predict both extinction and speciation in temperate angiosperm lineages. *BMC Evol Biol* 10: 162.
71. Lanfear R, Ho SYW, Love D, Bromham L (2010a). Mutation rate is linked to diversification in birds. *Proc Natl Acad Sci USA* 107: 20423–20428.
72. Lanfear R, Thomas JA, Welch JJ, Brey T, Bromham L (2007). Metabolic rate does not calibrate the molecular clock. *Proc Natl Acad Sci USA* 104: 15388–15393.

73. Lanfear R, Welch JJ, Bromham L (2010b). Watching the clock: studying variation in rates of molecular evolution between species. *Trends Ecol Evol*25: 495–503.
74. Langerhans RB, Gifford ME, Joseph EO (2007). Ecological speciation in Gambusia fishes. *Evolution* 61: 2056–2074.
75. Losos JB, Schluter D (2000). Analysis of an evolutionary species-area relationship. *Nature* 408: 847–850.
76. Losos JB (2008). Phylogenetic niche conservatism, phylogenetic signal and the relationship between phylogenetic relatedness and ecological similarity among species. *Ecol Lett* 11: 995–1003.
77. Macarthur RH, Wilson EO (1967) *The Theory of Island Biogeography*. Princeton University Press: Princeton, NJ.
78. Martin PR, Bonier F, Tewksbury JJ (2007). Revisiting Jablonski (1993): cladogenesis and range expansion explain latitudinal variation in taxonomic richness. *J Evol Biol* 20: 930–936.
79. Martin PR, Mckay JK (2004). Latitudinal variation in genetic divergence of populations and the potential for future speciation. *Evolution* 58: 938–945.
80. Maurer G, Portugal SJ, Cassey P (2011). Review: an embryo's eye view of avian eggshell pigmentation. *J Avian Biol* 42: 494–504.
81. Mayhew PJ, Jenkins GB, Benton TG (2008). A long-term association between global temperature and biodiversity, origination and extinction in the fossil record. *Proc R Soc B Biol Sci* 275: 47–53.
82. Mittelbach GG, Schemske DW, Cornell HV, Allen AP, Brown JM, Bush MB *et al*(2007). Evolution and the latitudinal diversity gradient: speciation, extinction and biogeography. *Ecol Lett* 10: 315–331.
83. Morlon HLN, Parsons TL, Plotkin JB (2011). Reconciling molecular phylogenies with the fossil record. *Proc Natl Acad Sci USA* 108: 16327–16332.
84. Nabholz B, Glemin S, Galtier N (2008). Strong variations of mitochondrial mutation rate across mammals—the longevity hypothesis. *Mol Biol Evol* 25: 120–130.
85. Nabholz B, Glemin S, Galtier N (2009). The erratic mitochondrial clock: variations of mutation rate, not population size, affect mtDNA diversity across birds and mammals. *BMC Evol Biol* 9: 54.

86. Nee S (2004). Extinct meets extant: simple models in paleontology and molecular phylogenetics. *Paleobiology* 30: 172–178.
87. Nikolaev SI, Montoya-Burgos JI, Popadin K, Parand L, Margulies EH, Antonarkis SE *et al* (2007). Life-history traits drive the evolutionary rates of mammalian coding and noncoding genomic elements. *Proc Natl Acad Sci USA*104: 20443–20448.
88. Nosil P, Flaxman SM (2011). Conditions for mutation-order speciation. *Proc R Soc B Biol Sci* 278: 399–407.
89. Ohta T (1992). The Nearly Neutral Theory of Molecular Evolution. *Annu Rev Ecol Syst* 23: 263–286.
90. Pagel M, Venditti C, Meade A (2006). Large punctuational contribution of speciation to evolutionary divergence at the molecular level. *Science* 314: 119–121.
91. Petit N, Barbadilla A (2009). Selection efficiency and effective population size in Drosophila species. *J Evol Biol* 22: 515–526.
92. Pierron D, Chang I, Arachiche A, Heiske M, Thomas O, Borlin M *et al* (2011). Mutation rate switch inside Eurasian mitochondrial haplogroups: impact of selection and consequences for dating settlement in Europe. *PLoS One* 6: e21543.
93. Purvis A, Hector A (2000). Getting the measure of biodiversity. *Nature* 405: 212–219.
94. Quental TB, Marshall CR (2010). Diversity dynamics: molecular phylogenies need the fossil record. *Trends Ecol Evol* 25: 434–441.
95. Rabosky DL (2010). Extinction rates should not be estimated from molecular phylogenies. *Evolution* 64: 1816–1824.
96. Rabosky DL, Lovette IJ (2008). Explosive evolutionary radiations: decreasing speciation or increasing extinction through time? *Evolution* 62: 1866–1875.
97. Rastogi RP, Richa, Kumar A, Tyagi MB, Sinha RP (2010). Molecular mechanisms of ultraviolet radiation-induced DNA damage and repair. *J Nucleic Acids* 2010: 592980.
98. Riek A (2008). Relationship between field metabolic rate and body weight in mammals: effect of the study. *Journal of Zoology* 276: 187–194.
99. Rohde K (1992). Latitudinal gradients in species-diversity—the search for the primary cause. *Oikos* 65: 514–527.

100. Rosenzweig ML (1995) *Species Diversity in Space and Time*. Cambridge University Press: Cambridge, UK.
101. Roy K, Goldberg EE (2007). Origination, extinction, and dispersal: Integrative models for understanding present-day diversity gradients. *American Naturalist* 170: S71–S85.
102. Ruiz-Pesini E, Mishmar D, Brandon M, Procaccio V, Wallace DC (2004). Effects of purifying and adaptive selection on regional variation in human mtDNA. *Science* 303: 223–226.
103. Rundle HD, Nagel L, Boughman JW, Schluter D (2000). Natural selection and parallel speciation in sympatric sticklebacks. *Science* 287: 306–308.
104. Schluter D (2009). Evidence for ecological speciation and its alternative.*Science* 323: 737–741.
105. Smith SA, Donoghue MJ (2008). Rates of molecular evolution are linked to life history in flowering plants. *Science* 322: 86–89.
106. Stadler T (2011). Inferring speciation and extinction processes from extant species data. *Proc Natl Acad Sci* 108: 16145–16146.
107. Stebbins GL (1974) *Flowering Plants: Evolution above the Species Level*. The Belknap Press of Harvard University Press: Cambridge, Massachusetts.
108. Stoletzki N, Eyre-Walker A (2011). The positive correlation between dN/dS and dS in mammals is due to runs of adjacent substitutions. *Mol Biol Evol*28: 1371–1380.
109. Storch D (2003). Comment on 'Global Biodiversity, Biochemical Kinetics, and the Energetic-Equivalence Rule'. *Science* 299: 346.
110. Thomas JA, Welch JJ, Lanfear R, Bromham L (2010). A generation time effect on the rate of molecular evolution in invertebrates. *Mol Biol Evol* 27: 1173–1180.
111. Thornton JW (2004). Resurrecting ancient genes: experimental analysis of extinct molecules. *Nat Rev Genet* 5: 366–375.
112. Tobias JA, Bates JM, Hackett SJ, Seddon N (2008). Comment on 'The Latitudinal Gradient in Recent Speciation and Extinction Rates of Birds and Mammals'. *Science* 319: 901.
113. Valentine JW, Jablonski D, Krug AZ, Roy K (2008). Incumbency, diversity, and latitudinal gradients. *Paleobiology* 34: 169–178.

114. Venditti C, Pagel M (2010). Speciation as an active force in promoting genetic evolution. *Trends Ecol Evol* 25: 14–20.
115. Wallace AR (1876) *The geographical distribution of animals. With a study of the relations of living and extinct faunas as elucidating the past changes of the earth's surface.* Harper & brothers: New York.
116. Wallace DC (2005). A mitochondrial paradigm of metabolic and degenerative diseases, aging, and cancer: a dawn for evolutionary medicine. *Annu Rev Genet* 39: 359–407.
117. Webster AJ, Payne RJH, Pagel M (2003). Molecular phylogenies link rates of evolution and speciation. *Science* 301: 478–478.
118. Weir JT, Schluter D (2007). The latitudinal gradient in recent speciation and extinction rates of birds and mammals. *Science* 315: 1574–1576.
119. Weir JT, Schluter D (2011). Are rates of molecular evolution in mammals substantially accelerated in warmer environments? *Proc R Soc B Biol Sci*278: 1291–1293.
120. Welch J, Bininda-Emonds O, Bromham L (2008). Correlates of substitution rate variation in mammalian protein-coding sequences. *BMC Evol Biol* 8: 53.
121. Whittle CA, Johnston MO (2003). Broad-scale analysis contradicts the theory that generation time affects molecular evolutionary rates in plants. *J Mol Evol* 56: 223–233.
122. Wiens JJ, Graham CH (2005). Niche conservatism: integrating evolution, ecology, and conservation biology. *Ann Rev Ecol, Evol Syst* 36: 519–539.
123. Wilson Sayres MA, Makova KD (2011). Genome analyses substantiate male mutation bias in many species. *BioEssays* 33: 938–945.
124. Woolfit M, Bromham L (2005). Population size and molecular evolution on islands. *Proc R Soc B Biol Sci* 272: 2277–2282.
125. Wright DH (1983). Species-Energy Theory—an Extension of Species-Area Theory. *Oikos* 41: 496–506.
126. Wright S, Keeling J, Gillman L (2006). The road from Santa Rosalia: a faster tempo of evolution in tropical climates. *Proc Natl Acad Sci USA* 103: 7718–7722.

127. Wright S, Ross H, Jeanette Keeling D, Mcbride P, Gillman L (2011). Thermal energy and the rate of genetic evolution in marine fishes. *Evol Ecol* 25: 525–530.

128. Wright SD, Gillman LN, Ross HA, Keeling DJ (2009). Slower tempo of microevolution in island birds: implications for conservation biology. *Evolution*63: 2275–2287.

129. Wright SD, Gillman LN, Ross HA, Keeling JD (2010). Energy and the tempo of evolution in amphibians. *Global Ecol Biogeogr* 19: 733–740.

130. Zapata FA, Gaston KJ, Chown SL (2003). Mid-domain models of species richness gradients: assumptions, methods and evidence. *J Anim Ecol* 72: 677–690.

Citations

CHAPTER 1

Baoye Wei, Jinzhe Zhang, Changxu Pang, Hao Yu, Dongshu Guo, Hao Jiang, Mingxin Ding, Zhuoyao Chen, Qing Tao, Hongya Gu, Li-Jia Qu, and Genji Qin, The Molecular Mechanism of Sporocyteless/ Nozzle in Controlling Arabidopsis Ovule Development, doi:10.1038/ cr.2014.145.

CHAPTER 2

Lindell Bromham, Peter F Cowman, and Robert Lanfear, Parasitic Plants have Increased Rates of Molecular Evolution across all Three Genomes, doi: 10.1186/1471-2148-13-126.

CHAPTER 3

S Blair Hedges, Jaime E Blair, Maria L Venturi, and Jason L Shoe, A Molecular Timescale of Eukaryote Evolution and the Rise of Complex Multicellular Life, doi:10.1186/1471-2148-4-2.

CHAPTER 4

Ravi K Singh, Klaus Gase, Ian T Baldwin, and Shree P Pandey, Molecular Evolution and Diversification of the Argonaute Family of Proteins in Plants, doi:10.1186/s12870-014-0364-6.

CHAPTER 5

Rajesh P. Rastogi, Richa, Ashok Kumar, Madhu B. Tyagi, and Rajeshwar P. Sinha, "Molecular Mechanisms of Ultraviolet Radiation-Induced DNA Damage and Repair," Journal of Nucleic Acids, vol. 2010, Article ID 592980, 32 pages, 2010, doi:10.4061/2010/592980.

CHAPTER 6

David Duchene and Lindell Bromham, Rates of Molecular Evolution and Diversification in Plants: chloroplast Substitution Rates Correlate with Species-Richness in the Proteaceae, doi:10.1186/1471-2148-13-65.

CHAPTER 7

Kathleen M Kay, Justen B Whittall, and Scott A Hodges, A Survey of Nuclear Ribosomal Internal Transcribed Spacer Substitution Rates across Angiosperms: an Approximate Molecular Clock with Life History Effects, doi:10.1186/1471-2148-6-36.

CHAPTER 8

E J Dowle, M Morgan-Richards, and S A Trewick, Molecular Evolution and the Latitudinal Biodiversity Gradient, doi:10.1038/hdy.2013.4.

Index

A

B

C

N

P

R

S

T

U

W